ENVIRONMENTAL CONSCIOUSNESS

Memory and Narrative Series
Mary Chamberlain and Selma Leydesdorff, series editors

Trauma: Life Stories of Survivors
Kim Lacy Rogers and Selma Leydesdorff, editors
(with Graham Dawson)

Commemorating War: The Politics of Memory
Timothy G. Ashplant, Graham Dawson, and Michael Roper, editors

Environmental Consciousness:
The Roots of a New Political Agenda
Stephen Hussey and Paul Thompson, editors

Memory and Memorials:
From the French Revolution to World War One
Mathew Campbell, Jacqueline M. Labbe,
and Sally Shuttleworth, editors

The Stasi Files Unveiled:
Guilt and Compliance in a Unified Germany
Barbara Miller

The Uses of Narrative:
Explorations in Sociology, Psychology, and Cultural Studies
Molly Andrews, Shelley Day Sclater, Corinne Squire, and Amal Treacher,
editors

Narrative and Genre: Contexts and Types of Communication
Mary Chamberlain and Paul Thompson, editors

The Clash of Economic Cultures:
Japanese Bankers in the City of London
Junko Sakai

THE ROOTS OF A
NEW POLITICAL AGENDA

ENVIRONMENTAL CONSCIOUSNESS

STEPHEN HUSSEY
PAUL THOMPSON
EDITORS

Memory and Narrative Series

TRANSACTION PUBLISHERS
NEW BRUNSWICK (U.S.A.) AND LONDON (U.K.)

First paperback edition published in 2004 by Transaction Publishers, New Brunswick, New Jersey, by arrangement with Routledge.

This book is printed on acid-free paper that meets the American National Standard for Permanence of Paper for Printed Library Materials.

Library of Congress Catalog Number: 2004044937
ISBN: 0-7658-0814-5
Printed in the United States of America

Library of Congress Cataloging-in-Publication Data

Environmental consciousness : the roots of a new political agenda / Stephen Hussey and Paul Thompson, editors.
p. cm. — (Memory & narrative series)
Originally published: London ; New York : Routledge, 2000.
Includes bibliographical references and index.
ISBN 0-7658-0814-5 (pbk. : alk. paper)
1. Environmentalism. I. Hussey, Stephen, 1967- II. Thompson, Paul Richard, 1935- III. Memory and narrative.

GE195.R66 2004
363.7'05—dc22

2004044937

CONTENTS

NOTES ON CONTRIBUTORS

Olivia Bennett is director of the Oral Testimony Programme at the Panos Institute, London. Since 1992 she has been working with partner organizations in more than twenty countries in Africa, Asia and South America, gathering and disseminating testimonies on development themes. She was managing editor of *At the Desert's Edge: Oral Histories from the Sahel*; contributing editor to Hugo Slim and Paul Thompson, *Listening for a Change: Oral Testimony and Development*; and co-editor of *Arms to Fight, Arms to Protect: Women Speak Out about Conflict*.

Giovanni Contini was formerly a research fellow at King's College, Cambridge, studying shopfloor bargaining, and is now responsible for the audio-visual archives of Tuscany. His publications include books on oral history, the Galileo engineering factory in Florence, and the marble quarrymen of Carrara and the landscapes they carve from the mountains, *Paesaggi del marmo*.

Timothy Edmunds completed his dissertation on environmental movements in Kazakstan at the Migration and Ethnicity Research Centre at the University of Sheffield. His research interests concern democratization and ethnic relations in the post-Soviet Communist environment, and he is now a Research Fellow at the Defence Studies Department of King's College London, working on civil–military relations in Central and Eastern Europe.

David W. Forrest wrote his doctoral dissertation in anthropology at the University of Miami. His research on the Maní has built on his long-time interests in the landscapes, water and soil resources, and the people and places of Mexico.

Stephen Hussey is a Research Associate at the School of Education, University of Cambridge. His previous publications include *Childhood in Question* and *Horrid Lights*.

Jaclyn Jeffrey teaches anthropology at the Texas A&M International University and also directs a cultural museum in Laredo. She has researched and published on ways that intangibles, like community memory and attachment to place, can be understood and used to help mitigate the negative impact of disasters in a community.

Daniela Koleva teaches the sociology of culture and oral history at the St Kliment Ohridski University of Sofia, Bulgaria. She has published a monograph on Pitirum Sorokin, *Sorokin: Ideas in the Theory of Culture*, and papers on the life history method and the oral history of state socialism.

Selma Leydesdorff is Professor of Women's History and head of the Belle van Zuylen Institute at the University of Amsterdam. In addition to her work on ecological disaster, she has written on Jewish history, including *We Lived with Dignity*.

Niamh Moore is teaching women's studies at the University of Sussex, and completing a doctorate on the relationship between ecofeminism and feminism. This work is based on life stories which she recorded with women campaigning against deforestation in Clayoquot Sound on the west coast of Canada. Her previous work on a proposed goldmine has been published as *The Omagh Goldmine: A Report on the Controversy*.

Bill Schwarz is the author of *The Expansion of England: Race, Ethnicity and Cultural History* and co-editor (with Mary Langan) of *Crises in the British State, 1880–1930*.

Paul Thompson is Research Professor at the University of Essex; senior research fellow, Institute of Community Studies, London. He is founder, National Life Story Collection, British Library National Sound Archive, and founding editor of *Oral History*. His previous books include *The Voice of the Past* and *The Edwardians*.

Jeff Wallace is Subject Leader in Literature at the University of Glamorgan. He has co-edited *Raymond Williams Now: Knowledge, Limits and the Future* and *Charles Darwin's Origin of Species: New Interdisciplinary Essays,* and has written widely on literature and science.

Andréa Zhouri studied anthropology at Unicamp University, Campinas, Brazil and researched for her doctorate on British activists in the Amazon at the University of Essex. She is now a lecturer in anthropology at the Federal University of Minas Gerais, Belo Horizonte, and is campaigning against dam-building in the mountains of Minas.

Introduction: the roots of environmental consciousness

Popular tradition and personal experience

Stephen Hussey and Paul Thompson

One of the most profound changes in human consciousness over the last fifty years has been the growing realization that nature is not an inexhaustible resource which we can plunder indefinitely to satisfy our immediate needs, but a complex and delicately balanced set of systems which can be fundamentally unbalanced or destroyed by human intervention. Consequently the entire future of human life depends on the safeguarding and evolving of environmental systems. But thinking about nature in this way cuts across not only widely held conceptions of traditional rural attitudes in earlier periods but also the short-term materialism of contemporary urban life. It has put a new set of issues on the political agenda, which belong easily neither to the left nor to the right. Crucial points along this path towards a greater environmental awareness were at first marked especially by the publication of powerful books, notably including Rachel Carson's *The Silent Spring* and also, ten years later, Meadows and Meadows' much criticized but undoubtedly catalytic report on *The Limits to Growth*,[1] which confirmed the long-term ecological impact of human industrial and population growth as a global political problematic. But subsequently, equally important have been direct signs confirming earlier warnings, and particularly the fears spread by events such as the Russian nuclear station catastrophe at Chernobyl, the discovery of thinning in the ozone layer and, most recently, the increasing evidence of climatic warming and disturbance. At the same time in many parts of the western world there has been a notable growth of green politics, ranging from conservative groups trying to ensure the preservation of natural rural heritage and the traditional countryside, through former peace activists now organizing mahogany boycotts in defence of the Amazon tropical forest, to tree-dwelling communities of young militants attempting through their defiance to save woods in the path of motorways. In Britain alone conservation groups have more members than the churches. Since the high peak of feminism in the 1970s, it is above all to environmentalism that idealistic young people in the west have become attracted, and to campaigning not only in their own countries but also to working for international NGOs in many parts of Africa, Latin America and Asia. This volume is a response to this context. Drawing on evidence both from personal memory and from intergenerational tradition, we wanted to explore a double issue: what is it in people's lives which gives them a consciousness of the changing environment, and what leads a minority into environmental activism?

This is clearly a crucial theme, for the future as well as about the past, and a large one, which we can only help to open up through offering new perspectives. Because the environment provides the context for all human lives, and affects every aspect of our lives, from eating and breathing to working and love-making, there are no clear limits to environmental and ecological literature: on the contrary, its scope seems infinite. It is the staple of botany and the bedrock of geography. There is a rich treatment of the theme in literary scholarship, including some significant written personal narratives. Political scientists and philosophers have thoroughly analysed the size, structures and divergent ideologies of green movements. Anthropologists and ethnologists have long concerned themselves with both the symbolic and the practical relationships between people and their environments. We could not hope to survey this vast literature here, even if we confined our attention to works in our own language, English. The books which we refer to in this introduction should be seen as a personal perspective on an impossibly wide topic.

Wide topic though it be, the environment represents, nevertheless, a surprisingly recent concern both for historians of all kinds, including oral historians, and for life story sociologists. This volume should be seen in part as a demonstration of the potential for historical and social research of this kind in many different directions. But oral narratives have been, in fact, surprisingly rare in this vast literature as a whole.

Of these many strands of writing, one of those which seems to us most salutary is literature, for it brings home the extent to which recent western attitudes to nature and the environment have deep and ancient roots, which can be traced back even to classical times. Animals and plants are motifs in the earliest European art, and you can find Roman vegetarianism in Ovid. Perhaps for a modern sense of environmental degradation we have to wait until the early modern period, but here certainly a striking instance is John Evelyn's book *Fumifugium* (1661), which is a vigorous tirade against the 'smoakes' of London and the city's 'fuliginous and filthy vapour' as a cause of widespread ill-health from coughs to consumption. And from this time onwards, as Keith Thomas has so vividly shown in his *Man and the Natural World* (1983), sensibility towards both animals and plants, and the imaginative projection of dimensions of human feeling into other living beings, rapidly intensified. This change of attitudes was the context for the well-known concern for nature of the romantic poets from the end of the eighteenth century. Thus when he wrote his famous lines in 'Auguries of Innocence' against the maltreatment of birds and animals, 'A robin redbreast in a cage/Puts all Heaven in a rage', William Blake spoke not only for himself, but for a growing number in his time.

From the romantic poets onwards there is clearly an unbroken current in English-speaking writing leading towards contemporary environmentalist thinking. Two of the poets are especially pivotal figures: John Clare and William Wordsworth. Clare is particularly remarkable from a social viewpoint, for he was the son of a village farm labourer born in 1793 in the village of Helpston, just above the edge of the Peterborough fens, and he can be seen as one of the first to express the views of the environment of the rural poor. John Barrell in his notable critical study, *The Idea of Landscape and the Sense of Place, 1730–1840*[2] draws out in particular the distinctions between upper-class appreciation of the landscape and that of Clare. The aristocrat had learnt, through the influence of the Grand Tour and Italian painting, to appreciate the wide view, but well shaped, in the painterly fashion followed by landscape gardeners; he felt uneasy at the 'formless' openness of open fields with their wide tracts of hedgeless space. Clare, by contrast, had a much more immediate, localized view, which he frequently noted in great detail, and he suffered when he saw even small changes to the familiar landscape around him: 'How oft I've sighed at alterations made' – 'A tree beheaded, or a bush destroyed'. Particularly strikingly, he disliked the reforming of his village landscape introduced by enclosure in Helpston when he was in his twenties, writing about it in a manner which reminds one of the New England poetry of Robert Frost in the mid-twentieth century, but with a reverse moral:

> Fence now meets fence in owners' little bounds
> Of field and meadow large as garden grounds
> In little parcels little minds to please
> With men and flocks imprisoned ill at ease.

Ironically it has been precisely the disappearance of such 'little parcels' of land 'imprisoned' by hedges and fences, and the return to earlier open 'prairie' landscapes, which has so distressed contemporary conservationists. Yet this in itself brings home the closeness with which environmental concern is related to personal memory. With Clare, close attachment to a particular place, memory of a local landscape known since childhood, 'made up' his sense of self, his very 'being'.[3]

The connection between personal memory and an intensity of feelings about the natural environment is still better known from the writings of Wordsworth, and because his poetry has long been recognized as a major landmark in this sense, it has been the subject of generations of critical discussion. Wordsworth has most recently been one of the writers singled out for attention by the new school of 'ecocriticism', which Jeff Wallace reviews in this volume: for he is the subject of Jonathan Bate's *Romantic Ecology* (1991). One does not, however, have to go along with the suggestion that Wordsworth was a poet of nature with no sense of place to recognize his role in highlighting the links between the contemplation of landscape and nature on the one hand and memories of childhood being and inner reflection on the other. These same connections recur in many subsequent authors. In the nineteenth century perhaps the most important instances are Ruskin and Morris, both of whom emphasized their early memories, Ruskin of the Lakeland crags which prepared him for the Alps, Morris of the Essex river marshes: 'I was always a lover of the sad lowland country.' Ruskin went on to become perhaps the greatest word-painter of landscape in English literature; Morris one of the first major English environmental activists, campaigning for the saving of woods, moors, country footpaths and commons from development and destruction. Later on, the geographical scope increasingly widens, perhaps most powerfully of all in Joseph Conrad's evocation of the tropics in *Heart of Darkness* (Massachusetts, 1902), a chilling tale told in autobiographical mode. And even W. H. Hudson, one of the best-known English naturalist writers of the early twentieth century, and perhaps most famous of all for his autobiographical memoir of *A Shepherd's Life* (London, 1910) on the Wiltshire downs, gave much of his energies to writing about Patagonian wildlife.

There are, of course, parallel currents in other literatures which may or may not be linked to the development of environmental consciousness and action in the regions from which they come: the North American musing on the forests from Henry Thoreau's *Walden* (Boston, 1854) onwards, the romantic influences of Herder or Goethe in Germany, or Rousseau's espousal of the wild and Lamartine's poetry in France.

There was no such early establishment among English-speaking historians of a tradition of writing about landscape and the environment, even though in France Michelet, himself one of the pioneers of oral history, wrote as a historical geographer, and some of the leading historians of the *Annales* school literally believed that fieldwork should include tramping the countryside to sharpen the researcher's sensitivity towards earlier peasant experience. In practice, however, environmental and landscape history is very much a development of the later twentieth century, and especially of the last dozen years.[4] Historians share the field with both anthropologists and historical geographers. And there are also interesting contrasts between historical approaches in different countries.

Thus in Africa, researchers are as likely to be social activists as historians: Melissa Leach and James Fairhead began their work (see below) in the field of forest conservation. Similarly, Hugh Brody's *Maps and Dreams* (1986), which is one of the most painstaking evaluations anywhere of the historical reliability of oral evidence, was the outcome of a campaign to document and protect Indian hunting rights in the Canadian far north. Olivia Bennett focuses her research primarily on development issues. As her chapter here admirably illustrates, collecting oral memories from villagers of ecological change and agricultural techniques, which are not available in written documentation, has become an important element in devising strategies for survival in particularly harsh environments. In North America, on the other hand, the focus has been on the much more fully documented national and local attitudes to the untamed natural environment, no doubt reflecting the longstanding American myth of the 'New World' and the wilderness frontier. Two notable instances are Max Oelschlager's *The Idea of Wilderness* (1991) and William Beinart and Peter Coates' *Environment and History: the Taming of Nature* (1995), which makes a comparison between North America and southern Africa. And in *Rescue the Earth* (1990), one of the very few published accounts of environmental campaigners based on life story interviews, edited by the Canadian veteran Farley Mowat who has been writing and struggling on these issues since the 1950s, the focus is again strongly on untamed nature, wildlife and wilderness. Mowat's interviewees are mostly top people in Canadian pressure groups. Michael Ninan's *People of the Rainbow* (1997), another rare oral history study, points in the same direction: it is about a loose anarchistic group, 'the Rainbow Family of Living Light', a mixture of environmentalists, feminists, peace activists and anarchists, who are drawn together by one common desire – to gather and celebrate in the remotest depths of the forests. One important exception to this pattern, however, is the important work of Kai Erikson on environmental disasters, beginning with his study, which again draws on oral testimony, of the collapse of a man-made dam, *In the Wake of the Flood* (1979).

The starting point for British environmentalist history, by contrast, has been much more the farmed countryside. The outstanding pioneering work of W. G. Hoskins, *The Making of the English Landscape* (1955), who first opened the eyes of many to see local landscapes as historical palimpsests created by generations over time, set out to interpret the cultivated landscape rather than nature. Hoskins, like John Clare, was particularly concerned by the changes brought about through enclosure in midland England. It has also been argued that he created a vision of rural England which offers a refuge for the writer from the present.[5] Similarly, Keith Thomas in his classic *Man and the Natural World* focuses on how, in the seventeenth and eighteenth centuries, civilized human feeling came to be imputed to nature. And similarly, the nearest British equivalent to environmental history is closer to an economic history of the human impact on nature in Britain, documenting the air and water pollutions which have resulted from the industrial revolution.[6] On the British environmentalist movement, there is an overview study for the late nineteenth century,[7] but none subsequently. Of more narrowly focused studies – which Linda Merricks characterizes as a 'very English approach' – the two most notable are on the history of nature conservation[8] and on rural conservation;[9] but there are still no important studies drawing on personal testimonies.

Paradoxically the absence of life story and oral history work in published environmentalist history is brought home by the work of two scholars who hold out greater promise by signalling even in their titles an ambition to link personal memory with landscape and sense of place: Schama and Nora. Simon Schama's *Landscape and Memory* (1995) – reviewed here by Bill Schwarz – is a thick, fully illustrated book full of brilliant passages, particularly about war and persecution in Italy and Eastern Europe. The opening section of the book is the most telling, set in the once-great forest of Bialowieza on the borders between Poland and Lithuania, and unravelling the murder and elimination of the Jews who were up until the Second World War so numerous in its villages. This was the region from which Schama's own family had escaped. There are especially telling moments when he clearly draws on personal or family memory to write about particular places: about the Thames mud and pebbles of his childhood; or the moment of discovery of a lost Jewish cemetery in the countryside. If only he could have drawn more reflectively on this vein. But unfortunately these are rare moments. As a whole, the book is astonishingly lacking in discussion of either the nature of landscape or of memory, or still more crucially, of how the two are connected. Schama never attempts a visual analysis of landscape and, essentially, landscapes are simply described as stage scenery for events; while memory is too diffuse to identify as a factor, and in fact most of the book is written from ordinary historical sources. It is a great book to dip into but despite its beguiling title it does not offer any answers to the questions from which we started.

Pierre Nora comes from a more solemn French tradition of scholarship, researching in the context of the *Annales* school, the long-time perspective of the *longue durée* and the total economic and cultural history of Braudel. Behind that, of course, lies the nineteenth-century French sensitivity to cultivated and industrial landscape expressed by novelists like Zola, or in the painting of the Impressionists. Oral history is relatively rare in France, so it is less surprising that Nora does not use it as one of his sources. His multi-volume *Les Lieux de mémoire*, translated into English as *Realms of Memory* (New York, 1996), therefore holds out promise of a different perspective on the environment. In fact we find a remarkable medley of themes, ranging from the medieval cathedral, pedagogical manuals for history and geography teachers, Marcel Proust, and war memorials, to the Académie Française, the Tour de France bicycle race and local gastronomy. It begins to look as if anything cultural from the past is a form of memory. As for the role of places in memory, there is never a satisfactory definition of *lieux de mémoire*, so that it is no accident that the English translation substitutes 'realm' for 'place'. Some of the contributors use the phrase 'mnemonic sites', which if examined more particularly could have been illuminating. Once again, when opened the package does not bring the new approach of which its title holds out hope.

Thus despite the growth of research on the history of the environment and on environmental movements, publications drawing on personal experience or collective memory have remained surprisingly rare. Our intention in this volume has been above all to highlight the potential fruitfulness of such an approach. For this reason we have tried to give it a double character, both diverse and interlinked. In terms of discipline we have chapters drawing on oral historical, developmental, political, sociological and anthropological approaches, and they concern change in five continents. There are also many connections between them. Two chapters are concerned with the former Communist republics of Bulgaria and Kazakstan. Three others focus on change and campaigning in the central Americas from the Texas border to the Amazon, but the chapter on the Amazon leads straight back to environmental activism in Western Europe. Trees are a recurrent theme. Another linking theme, in the chapters on both Africa and Western Europe, is the nature of peasant consciousness of the environment, and how ancient in origin some of our own beliefs and traditions may be. And all these chapters contribute in their different ways to our growing understanding of the roots of environmental consciousness, and some also to the sources of militancy.

We begin with Europe, our own backyard. In writing of 'The English, the trees, the wild and the green', Paul Thompson takes a wide canvas, tracing particular traditions and attitudes to trees and forests back to prehistory. It emerges that there are indeed very longstanding traditions, such as of the wild people of the forests, which have continued to influence generation after generation up to the present – including even some respected anthropologists. Yet on the other hand the linking of the green man with the Druids – a connection of which a lecturer at the newly created Clun Green Man Festival in Shropshire maintained we all carry genetic memories in our DNA – turns out to be an imaginative reinvention of tradition. It is with the evolution and transformations of these traditions over centuries, and how they provide one source, often unconsciously, for contemporary attitudes, that this chapter is concerned.

The English today are passionately involved in the preservation of trees and forests, both locally and internationally. At home the booming of the Woodland Trust to its present 200,000 members, taking a new wood every week into protection, is one indication of the strength of this cause; another is the militant campaigning by young activists in the paths of runways and motorways, defying the bailiffs and police from their underground tunnels and treetop encampments. But equally, as the later chapter by Zhouri illustrates, the English have thrown themselves into campaigns to protect the tropical rainforests in Brazil and elsewhere. Yet paradoxically the English have also been ruthlessly destructive of trees, and for centuries: four fifths of the forest cover of England was already felled by the early middle ages, and today it is the least in Europe. Even after 1945, thousands of miles of hedgerow trees were destroyed by farmers, while East African tropical forests were being felled to make fields for groundnuts as a consequence of systematic British colonial economic policy. Perhaps the English love of trees, which as early as the seventeenth century was asserted by John Aubrey to stem from Druidic tradition, needed the support of traditional images precisely because the lessons of practice were so negative. In the same way the fight to save the tropical forests of the Amazon has been partly based on reformulated old myths: from the Elizabethan venturers' belief in hidden fabulous metallic wealth which Raleigh sailed to seek in El Dorado, to the chemical source of human wellbeing offered by the Victorian Dr Palmoil, to the contemporary vision of the forests as a vast lung essential to the workings of the global climate and so to human survival.

Trees are of course of major symbolic significance for many cultures besides the English. Indeed, Laura Rival's recent anthropological collection on

The Social Life of Trees (1998) encompasses coconuts in Pacific fertility and funeral ceremonies, sacred jackfruit groves in India, First Nations fighters for the Canadian forests, Japanese ritual family woods, and western grassroots campaigns for the tropical rainforest. The cultural importance of trees is equally diverse historically. Foliage is an important motif in Persian and in Byzantine art. There are sculptured leaves in most of the great cathedrals of Europe. The Tree of Jesse, illustrating the genealogy of the Holy Family, is a widespread Christian image, in Europe and beyond; still more so its secular equivalent, the family tree. The Tree of Life is a favourite image of Mexican potters. There are Indian legends of trees that speak, and one of the most successful movements for forest protection has been Indian, the Gandhian style Chipko movement of treehugging women (among them Sudesha Devi, whom Olivia Bennett quotes in her chapter). Thompson however focuses more specifically on two particular traditions out of many concerning trees and forests: the fears and fantasies about the strange people who lived in the forests, as symbolized by the green man and the wild people. It needs to be stressed that neither of these traditions has in fact been peculiarly English: both are found right across Western Europe from Germany to Spain; and the image of the wild people also followed the European conquerors to the New World.

Of the two images, that of the green man, the face with branches growing from his mouth, although much better known today, is also the more mysterious. Surprisingly, there are no oral traditions of green men and there is no clear sign that they were part of popular culture in the past. The medieval green man is a puzzling case of a transgenerational and international symbol which was expressed only visually, as a mythical image, a mute narration through form rather than words. It is just possible that in Devon, where the images are particularly prevalent, they do have a connection with the Mayday King and Queen. But the most likely explanation of their meaning is that green men were a symbol of the masons, and for them they may well have represented a particular kind of fertility, that of artistic creation. Certainly they disappear from the sixteenth century, when the earlier culture of the masons, centred on the great churches where they congregated to work and left most of these green men images, dissolved.

The contemporary cult of the green man was a reinvention of the folklorists of the late nineteenth and early twentieth century. First in *The Golden Bough* (1890) Sir James Frazer identified Jack the Green, the Mayday dancer in a green bush, as a symbol of the sacred tree of rebirth. Next, traditional dancing sides were discovered by Cecil Sharp and others, and given a central position in English folklore tradition. Finally, in an article published in 1939 by Lady Raglan, the medieval heads were linked with the Mayday dancers and the sacred tree. It is these associations which the green man carries today, as an ancient symbol of fertility and rebirth going back to the Druidical roots of English culture.

The wild people, by contrast, have a much more certainly known past, for they can be found not only as visual images but also in popular oral tradition, and they can be clearly traced back to biblical and Babylonian epics. Like green men, medieval wild people were associated with fertility, but in a darker, more animal way, feared as seducers of wives and children. However, their lusty sexuality was not only seen as dangerous but also envied, and becomes in many representations utopian, frolicking and lovemaking in the open air. Feeding off forest roots, with their long-haired women riding deer bareback, the wild people were also admired for their closeness to nature and to animals and their command of them.

These contradictory images became increasingly separated. With the Conquest of the Americas and the expansion of Europe, the subjugated natives of the colonies were often viewed as monstrous races manifesting all the worst aspects of wild people, and hence needing control and conversion. But there were other strands. William Morris' advocacy of woodland camping and his prose romances about idealized forest societies led through to the Woodcraft Folk movement in North America and Britain. There were tropical romances too. Captain Cook and his crew found a sexual paradise in the Pacific islands. Gauguin painted the same utopian innocence in Tahiti, and Margaret Mead sought to document it in Samoa. And the same utopian tradition of the wild people underlies some of the surprising writings about the forest peoples of the Amazon, whom Theodore Roosevelt likened to a vision of Adam and Eve before the Fall. Especially strikingly, the noted Brazilian anthropologist Marcos Santilli has described one isolated forest group as living in self-sufficiency, without power or oppression, free of madness and sexual conflict, and living in 'coexistence with wood, adaptation to nature, co-operation, spiritual evolution': truly a last utopia of wild people.[10]

These traditions of the strange people of the forest, in short, are still alive, whether explicitly recognized or not, and contributing to our consciousness of the environment and the relationship between human beings and nature. There are indeed – as we shall see, for example, from Africa – instances of peasant cultures living in a positive symbiotic relationship with nature. But equally, peasants can be exploitative and ruthless in their attitudes to the natural world, as Giovanni Contini brings out in his chapter, 'Animals, children and peasants in Tuscany'. Here in the small village of San Gersolé, close to Florence and in the sharecropping region, a far-sighted local teacher, Maria Maltoni, from the 1930s to the 1950s instructed her children to keep journals of their everyday lives, including relationships within the family and with animals and nature. These journals, which read almost like transcribed oral documents, rudely contradict the widespread Italian view of a peasantry living in a mutually harmonious relationship with nature. Family relationships as portrayed are bad enough, hierarchical and violent, men controlling women, and adults subjugating children with fists and whipping. But children in turn wreaked a similar violence on animals and nature. They wantonly killed birds, cats, porcupines and snakes, they ransacked birds' nests and they pulled insects apart.

Interestingly, this destructive delight in killing could be combined with a remarkable knowledge of different species. This must have been one aspect of peasant culture which misled outsiders. But commentators also projected what they hoped to find. Thus when Maria Maltoni published a selection of drawings of flowers and fruit by the children, their perfection was seized upon by Italian intellectuals such as Emilio Cecchi and Italo Calvino as evidence of the closeness of peasants to nature: how else could semi-literate youth paint so wonderfully? In fact, as Contini's oral interviews have revealed, the drawings were far from spontaneous. On the contrary they were the outcome of fierce coaching by Maria Maltoni, to the point of biting pupils who produced unsatisfactory work – a style of teaching which neatly echoed that between local parents and children, or peasants and nature.

Daniela Koleva in her chapter 'Narrating nature' examines the forms of remembering among generations of Bulgarians who have lived through its transformation from a peasant to first a Communist and then a market society. She did not deliberately set out to collect environmental narratives, but nevertheless found that they occurred in nearly a third of her interviews. She suggests that in general particular aspects of a life story are tied to particular environmental settings. When people tell their stories, changing environments are deployed to symbolize changes in life, while stable environments are the setting for stable identities. Quite often when asked for a date in their lives people will answer with a place. She finds a particularly strong contrast between the memories of those who have migrated, when movement gives a life story clear phases linked to different places, and those who have stayed in a single environment and tend to speak of an almost timeless past, often talking in a past continuous tense and fusing their own experience with those of others transgenerationally.

Koleva divides the memories which she has recorded into two different types, the prosaic and the romantic. Prosaic memories are typical of older former peasants still living in the countryside. For them, formerly time was measured by seasonal work, the day by sunrise and sunset. They saw the rivers, trees and hills around them not as scenery, but as something to be handled and dealt with. They speak especially of the details of farming and skilled work, and of their resentment when under Communist collectivization that they were bossed around by outsiders who often did not understand these skills. An interesting member of this group is a construction engineer who was responsible for many local irrigation works manipulating and changing the landscape. In retrospect – although this is far from the extreme damage brought to the environment in many Communist countries, including Kazakstan of which Timothy Edmunds writes here – he recognizes that some of this work was not effective in the way intended and brought unnecessary changes to the local environment.

Romantic attitudes were much more typical of city-dwellers. They see nature as harmonious and continuous, essentially 'beneficial' and they used adjectives for it such as 'pure', 'beautiful', 'healthy' and 'fresh'. There are often descriptions of weekend picnics, and mushroom and herb gathering in the 'virgin forest'. Often memories of important relationships are linked to love of the sea and the mountains. It is particularly with those whose childhoods were spent in the country before they moved to the town that life stories carry a nostalgic romanticization: childhood was 'a fairy tale', the river was full of fish, and the melons the sweetest ever eaten. The same people bemoan subsequent environmental destruction and 'dying nature'. Do they represent, perhaps, an especially fertile recruiting ground for environmental activisim?

However much human beings may wish to think of nature and the environment as essentially benign, there are nevertheless times of disaster – droughts and famines, hurricanes and floods – when this viewpoint ceases to be tenable. Kai Erikson has written of one such episode in his book, *In the Wake of the Flood*[11] and Selma Leydesdorff writes here of another, the Dutch North Sea floods of 1953; both draw especially on oral history evidence. With Erikson the flood at Buffalo Creek in the Appalachians, which left 4,000 people homeless, was blamed on the company which managed the dam which broke. People felt that to blame God for human failings would be blasphemous. In the Netherlands it became clear after the disaster, in which 1,800 lost their lives, that the dikes had been in a poor condition, and they have since been raised to a far higher level. But the people of Zeeland, as religious as their Appalachian fellow-sufferers, took a different view of God. They saw the floods as a punishment for their sins, and they remembered it in Biblical terms, as a 'deluge' which has left nothing unaltered.

Both in Zeeland and in the Appalachians, the legacy of environmental disaster has been a people living in fear of nature. Fishing communities have always lived in this spirit, and Leydesdorff points to another instance, the French farmers living below a mountain whose cliffs fell as long ago as the middle ages. As one Dutch survivor put it, 'I live here on the isle with the fear it will happen again . . . The wind howling through the trees keeps me awake.'

Natural disaster has struck still more severely at the rural communities of whom Olivia Bennett writes in her chapter, '"Our land is our only wealth"'. She draws on a series of oral history projects linked to relief and development work and carried out by Panos in the 1990s, beginning with their ambitious Sahel project for which 500 people were interviewed, mainly elderly, in eight sub-Saharan countries affected by climatic change in the form of prolonged drought. This is a very large region in which nature has always been problematic, and eking out a living from the marginal terrains of the Sahel stretched skills to the limits. A notable collection of their testimonies is published in Nigel Cross and Rhiannon Barker, *At the Desert's Edge* (1991). Many of the interviews from these projects include evidence of powerful feelings of attachment and reverence for nature, suggesting a peasantry who saw man and nature as spiritually as well as materially intertwined. Thus in the Sahel, land was seen as sacred, unsaleable. Similarly, in Ethiopia a village priest likens the need for fallow years to the land's need to rest like a human patient. He suggests the suffering of the land is their own: 'We are inextricably bound to her. Her poverty is our poverty, we suffer together.' Even changes in human knowledge are conceived of in relation to natural rather than social change: 'it seems as if the drought is drying up the skills and knowledge that we had'.

Bennett sees these interviews as especially important since they voice the rarely heard experiences of the marginalized people affected by development, who are the true 'experts' in its realities. Specialized local knowledge of an environment can also provide crucial clues for revitalizing it, identifying systems of water conservation and husbandry, favoured species, and also dangers. In Africa, perhaps more than anywhere else, indigenous peasant knowledge has been demeaned and disregarded by European colonists. One striking instance of this is provided by Melissa Leach and James Fairhead's study of Guinea, *Misreading the African Landscape* (1996). From their first occupation of the territory in the 1890s, the French had assumed that a dense humid forest originally covered the landscape of Guinea, and that its depredation was due to felling by local tribes. They used this interpretation as one argument for wresting control of the environment from locals, and imposing repressive policies, including the death penalty for starting a bush fire, to enforce their own view of land management. More recently, with deforestation blamed for the drying climate and world concern about the tropical forests, major international funding has been attracted to 'restore' Guinea's tropical forests. But Leach and Fairhead's impressive study, which draws on aerial photography, participant observation and oral history interviews, proves that the European interpretation has been completely wrong: so far from felling being associated with human habitation, as the settlements have advanced into the savannah their local peasant inhabitants have been planting belts of trees to surround them. They have been making the forest grow rather than shrink. In short, the French understood what was happening to the local environment, and how to look after it, much less well than the local peasantry.

The Panos Highland Communities project includes studies of small mountain communities in Peru, the Himalayas and Lesotho in southern Africa. The Indian Himalayan study is unusual in that here there was collective resistance to change, including both the Chipko movement to protect the forests and the militant resistance to the Tehri dam. Bennett also recounts some of the pathetic distress expressed by mountain peasants forced off their land, losing their fields and their family graves, to make way for the Mohale reservoir in Lesotho. As an old woman put it, 'It will remain as a rock in my heart when I think of the place that I am being removed from.' Particularly strikingly, they speak of their land not just as fields they have cultivated but as a specific place where their detailed knowledge makes sense: so that they knew where and when to find medicinal plants, raw materials for building or clothes, places for sheltering livestock or the first spring pastures. Resettlement meant the crucial loss of 'the wisdom of living in that place', and even if they were fortunate enough to get new fields, they would need once again to 'struggle to learn that land'.

Two other chapters concern adaptations by rural communities in Central America, one of which has been resettled while the other has remained in the same settlement for centuries. Jaclyn Jeffrey in 'Using community memory against the onslaught of development' writes of the Texan town of Zapata, settled by Spanish colonizers in the mid-eighteenth century, in the harsh arid thornbush country now the borderland with Mexico. Survival for the farmers and ranchers here still requires traditional ecological skills in monitoring water supply, size of herds and vegetation, such as knowing precisely when to burn the needles off the prickly pear so that the cattle can eat it. The whole of the small town was resettled in 1953 when a reservoir was built in the valley, resulting in the loss of many beautiful houses and also an influx of Anglos with different notions about how to exploit the environment. Jeffrey shows how there are distinctive Spanish- and English-speaking memories of the past and the town before it was inundated, and, linked to these memories, competing views of how the town should be developed now. Control so far has been maintained by the Spanish-speaking group, deliberately impeding the use of the new lake for tourism, and maintaining a sense of cohesion through a combination of their hold on local politics and their ability to promote their collective memory through the Zapata County Historical Commission.

David Forrest's 'Signs of things to come' is about the town of Maní in the Mexican Yucatán, which, by contrast, is a small but historic centre of the Maya, who have been here since long before the Spanish Conquest. The Maya had developed in the distant past a twofold system of agriculture, combining an irrigated infield for fruit trees and poultry with an outfield which was cut from the communal forest in a slash and burn rotation. Forrest got to understand their views especially through talking as they walked around their fields. He found that this talk ranged from practical discussions of the everyday use and control of environmental resources to Mayan cosmology and mythology: 'I was amazed at the richness of the narratives told to me.' The Maya, perhaps still in the shadow of the destruction of their ancient civilization which may have been linked to an episode of drought, hold a cyclical view of time and a pessimistic view of the environment, predicting future collapse. They have certainly had to survive very difficult episodes in the past, including plagues of locusts, and more traditional locals still use special forest altars for rain ceremonies. Most recently, from the 1960s there has been a government-encouraged move for the conversion of forest into more irrigated land for fruit and vegetable growing using fertilizers. When this new agriculture went into crisis in the mid-1990s through the doubling in price of fertilizers the traditional prophecies of doom were seen to be realized.

Maní again, in short, is another instance in which people draw their understanding of their local environment from long transgenerational traditions.

Such traditions can again be one important element in the shaping of a more radical consciousness of the environment which leads towards activism. The final three chapters of the volume look at different kinds of activism in different contexts.

In a chapter focusing on one of the former Soviet republics, Timothy Edmunds examines 'The environmental movement in Kazakstan'. It seems initially surprising that this movement has not only played a key role in developing a post-Communist civil society, but has also been among the most politically effective environmental movements in the world, having dramatically succeeded in its campaign against Soviet nuclear testing after both direct action and the collection of a million signatures to a petition, and is still the dominant political alignment in the republic rather than the nationalism which outsiders had predicted. There seem to be three important elements contributing to this remarkable mobilization: traditional Kazak identity, changing popular attitudes to Soviet environmental activity, and direct evidence of damage.

There is a traditional link between love of nature and Kazak national identity. The national flag includes an eagle and winged horses, and national tradition emphasizes a romantic vision of the nomadic pastoral life as symbiotic with nature. The north of the vast country remained pastureland for centuries, until much of it was destroyed by Khrushchev's ill-advised conversion to arable for his 'virgin lands' scheme in the 1950s. The Soviets, however, had a confusing attitude to the environment. On the one hand they had on paper strict laws for the protection of the environment, for example from industrial pollution through factory emissions; and it was not immediately obvious that these laws meant little in practice. On the other hand, they shared the view – which was certainly not exclusive to Communists in the 1930s and 1940s – that nature could readily be manipulated for human benefit. Throughout the Soviet Union ambitious schemes were initiated to improve and expand the land under agriculture, a transformation whose scale is perhaps only comparable with the later terracing of the mountainsides by Chinese communes. It seems likely, although it is very difficult to judge in retrospect, that there was initially a good deal of genuine popular enthusiasm for these massive environmental manipulations. One can find, for example, in an optimistic wartime survey of development in Soviet Asia, a chapter on Kazakstan which cites local poetry celebrating progress: 'Go on, for the covered ox-cart, forward! / There is no road back!'; 'Our Union is an Express, flying over bridges, / Past verdant meadows, across the broad acres. / Our land is well off, / Our people the stronger, / And free to move as the clouds . . .'[12] Still more striking is their report from Uzbekistan which describes the collective work in the 'digging of ditches that transforms deserts into blooming fields of cotton, melons, and even flowers': local farm families walk for miles, the men in long white shirts and the women in bright colours; the Young Pioneers march up with their banners and bands; there are field kitchens and native dancers performing; and regular crews instruct the volunteers, sometimes 100,000 in a day, so that 'literally thousands of men, women, and children swarm over the diggings, pitch in with hand shovels, and remove dirt by the basketful'.[13] No doubt the authors were in many ways misled; but it seems unlikely that there was not some genuine enthusiasm behind the generating of this massive environmental effort.

What crucially changed opinion was the growing evidence of the damage which Soviet policies were causing. In the Soviet Union as a whole the Chernobyl nuclear power station explosion in 1986 was a turning point, and one of its legacies was the spread in many parts of the Union of environmental groups, mostly shortlived and concerned with local issues such as factory emissions or power stations. These local movements were tolerated by the Communists partly because they seemed a safe outlet for protest. In Kazakstan, however, the movement developed in a much more powerful way, and there can be little doubt that one reason was simply that the environmental damage wrought by the Soviets was exceptionally severe. There was damage here as in so many other places from pollution by chemical fertilizers and by industrial factories. But this was nothing to the Aral Sea and the nuclear testing. The diversion of the rivers feeding the Aral Sea had been made in progressive hope, to provide new irrigated land for growing cotton. But the improvers had failed to take into account the consequences. The Aral Sea has now shrunk to half its size, its once flourishing fishing industry has vanished and its wildlife has been destroyed, duststorms from its exposed seabed are destroying the formerly good land on its shores, and the local climate has changed radically, shortening the growing season. In short, a scene of massive devastation has emerged. And the impact of hundreds of nuclear bomb explosions at Semipalatinsk, the Soviet Union's prime nuclear testing facility, was if anything worse still, for it resulted in a trebling of local cancer rates. But in that case at least something could be done. When Kazak environmentalism grew into a massive protest movement in the 1980s, it was led by an esteemed local poet, and its focus was on halting the nuclear testing. And the tests did cease.

Niamh Moore's 'Paths to ecofeminist activism' is not a study of a movement but a close-up of three women activists in the north-east of England, seeking to understand what has led each of them to their current environmentalist involvement. They have set up a local branch of the Women's Environmental Network and are active in campaigns about a local bypass, the reopening of a disused railway line and the recording of sacred sites. Personal and everyday experiences emerge as important in every case. To begin closest to home, Moore observes: 'Gardening seems to have many meanings for Karen. It is self expression, it is a document of a life story, it is ritual, it is political activism, it is evidence of the connection between the personal, political and spiritual in Karen's life.' It turns out that gardening has been important for all three women, often from childhood. There also seems to be a possible connection between their experiences of abuse from men and their view of nature as suffering abuse. All three have been influenced by higher education, and all three have been politically active in a variety of ways: as peace campaigners, local Green or Labour councillors, and so on. Karen especially has a passion for trees, and an important early experience was joining a protest camp against forest felling in western Canada. Their activism, in short, has in each case been stimulated by personal experiences which go back to their childhoods.

Such evidence from life stories supplies the accounts of personal experience which are a missing dimension of almost all critical studies of environmental movements hitherto. They need of course to be contextualized again in relation both to direct environmental experience – the bypass or the forest felling – and to the ideologies, including collective memory and myth, through which people interpret it. With the forest felling in Canada against which Karen protested, for example, it is no accident that the first great campaigner, Grey Owl, the articulate Indian who starred in films petting beavers and moose, whose lectures and books in the 1930s influenced thousands, and who was even privately received by the royal family, turned out on his death to have been living a romantic fantasy. He was not a half-caste North American Indian but an English immigrant from Hastings, Archie Belaney, who had been brought up on books such as those of Ernest Thompson Seton which romanticized the Indians of the northern wilderness and their closeness to nature; and in living out this myth he even married two Indian women bigamously. He was a fantasy wild man made famous.

Dreams have led environmental campaigners to many distant countries. In her 'Pathways to the Amazon', Andrea Zhouri examines what draws British campaigners to fight for the Brazilian rainforest. Interestingly she found these activists suspicious of her life story interview method, for they felt that this might portray them as emotional rather than present the rational public front which they wished. When asked about themselves, they tended to give their qualifications first. Nevertheless common earlier experiences did again emerge as important. In particular, almost all experienced travelling as part of the process of their political engagement. Some had travelled from childhood because a parent was a diplomat or in the military or a company employee in Latin America or Africa; others began travelling, and experienced their first taste of adventure, later on as adults. Beyond this, however, she distinguishes two groups of rainforest campaigners. The younger generation, in their twenties and thirties, believe in a political practice based in science and research, and this reflects their academic training which is typically in forestry, geography or ecology. They have no links with Latin American culture and people, and their focus is on the trees rather than the people of the forest. The middle generation, in their forties, are as much concerned with the forest peoples as the trees. They were trained in social science or literature, and they have much broader political experiences, sometimes including activism in the peace movement. In their travels, what most powerfully changed their consciousness was the direct experience of famine or poverty, and this stays with them as environmentalists: 'if you don't involve social justice as well – if you are just thinking about conserving a tree or a frog, and you don't look at the whole thing, then you are not going to be able to save it'. In short, as with Moore's ecofeminists, activism is stimulated by a range of personal experience, including travel, education and politics; and beyond that too, the special intensity of the Amazon rainforest as a symbol, and the real destruction which it is suffering.

It is precisely in its power to bring together public and private influences in the generation of environmental consciousness and activism that we believe an approach based on memory, in terms both of personal experience and life stories and also of collective memory, myth and tradition, has a unique potential. Clearly the chapters in this volume can be no more than a beginning. In every part of the world today there are campaigns about the environment, victims of development, and ordinary men and women who have seen the settings of their lives radically transformed. Only when the stories of far more of them have been heard will we begin to be surer about why some people care about their environment, and a minority care to the point of fighting for it, while for so many others it is never an important concern. We see this volume as one step on the path to that understanding: an understanding which we now know is vital for the future of humanity.

Let us conclude, in this spirit, by hearing a voice from an interview recorded by one of us earlier this year: the voice of a welder in an inner London computer parts factory, Selvin Green, born in Jamaica in 1937. Selvin was brought up by his grandmother in the beautiful mountain countryside of eastern Jamaica. But when as a teenager he moved into the city of Kingston he was delighted to turn his back on the country, seeing it as an escape from a monotonous fate:

> Oh, it was great! It was great! When I was in Kingston, one is like you're in heaven! Yes, it was great! Because I'm always at work . . . because I didn't want to work in the fields, in the cane fields . . . I hate that . . . Kingston is the big bright lights, and everything is there.

When he was 24 he travelled again, joining his mother who had gone ahead to London, getting chilblains that first grim English winter – 'it was cold. Man, it was so cold'. But by this time he was in a new phase of life, married with children, and he soon decided to move out of the metropolis to a smaller town to give his children a better chance and all of them 'fresh air'. Selvin is back in London now. But in the interim he has also spent eight years back working in Jamaica, and he is longing to retire there: 'Oh yes! Oh very yes! I'm just raring to go! Oh yes, I will go. If I'm not dead, I'm gone.'

Perhaps most striking is the lyrical way in which he now speaks of a countryside which once, as a teenager, he had been so glad to escape. He speaks of returning to 'my home' with the passion of a lover of nature. What draws him so powerfully is

> The lifestyle, the place, the beauty. The simplicity of life. I want to go out the back and pick a lettuce, just off the real land grown on. I want to pick orange. I want to hear the birds. I want to see the coconuts, I want to hear the wind blow between it . . . I want to see the bees fly up to the flowers, and I want to stand there, because I used to do that, and look at it taking the nectar from the flowers, you know?
>
> Cos I used to do that. I used to watch, I used to get disciplined for it. When I'm going down to the stream to get water . . . That's like you see a bird fly in the tree, and it make a sound, a whistle; when you look, you see another one fly, come along, and some communication going on between those two bird. These are the things I like.

In England Selvin has been a quiet labour activist. In Jamaica environmental activism is only recently beginning to make a mark. But from this recording of his memories it is easy to see how the twists of Selvin's life have transformed the youthful lover of the bright lights of the city, only too glad to escape the 'simplicity' of rural life, into a future returnee who may be just the kind of activist who Jamaica's nascent environmentalism needs. And his story can stand for countless different others. For humanity's future on this planet, we need to listen to them.

Notes

1 R. Carson, *The Silent Spring* (Boston, 1962); D. H. Meadows and D. Meadows, *The Limits to Growth* (New York, 1972).
2 J. Barrell, *The Idea of Landscape and the Sense of Place, 1730–1840: An Approach to the Poetry of John Clare* (Cambridge, 1972).
3 Barrell, *The Idea of Landscape*, 114, 143, 174.
4 For reviews, see L. Merricks, 'Environmental History', *Rural History*, 3 (1996), 97–109; M. Williams, 'The Relations of Environmental History and Historical Geography', *Journal of Historical Geography*, 20 (1994), 3–21; A. W. Crosby, 'The Past and Present of Environmental History', *American History Review*, 100 (1995), 1, 177–89.
5 D. Matless, 'One Man's England: W. G. Hoskins and the English Culture of Landscape', *Rural History*, 4(2) (1993).
6 B. W. Clapp, *An Environmental History of Britain since the Industrial Revolution* (London, 1994).
7 P. C. Gould, *Early Green Politics: Back to Nature, Back to Land and Socialism in Britain 1880–1900* (Brighton, 1988).
8 D. Evans, *A History of Nature Conservation in Britain* (London, 1991).
9 J. Sheail, *Rural Conservation in Inter-war Britain* (Oxford, 1981).
10 M. Santilli, *Àre* (São Paulo, 1987), 15–16.
11 Erikson's work draws on an American sociological research specialism in disasters, first developed in Washington in the 1950s.
12 R. A. Davies and A. Steiger, *Soviet Asia* (London, 1943), 62–3.
13 Davies and Steiger, *Soviet Asia*, 80.

Select bibliography

Barrell, J., *The Idea of Landscape and the Sense of Place, 1730–1840: An Approach to the Poetry of John Clare* (Cambridge, 1972).

Bate, J., *Romantic Ecology: Wordsworth and the Environmental Tradition* (London, 1991).

Beinart, W. and Coates, P., *Environment and History: The Taming of Nature in the USA and South Africa* (London, 1995).

Brody, H., *Maps and Dreams: Indians and the British Columbia Frontier* (London, 1986).

Carson, R., *The Silent Spring* (Boston, 1962).

Clapp, B. W., *An Environmental History of Britain since the Industrial Revolution* (London, 1994).

Crosby, A. W., 'The Past and Present of Environmental History', *American History Review*, 100 (1995), 1, 177–89.

Cross, N. and Barker, R., *At the Desert's Edge* (London, 1991).

Davies, R. A. and Steiger, A., *Soviet Asia* (London, 1943).

Erikson, K., *In the Wake of the Flood* (London, 1979).

Evans, D., *A History of Nature Conservation in Britain* (London, 1991).

Evelyn, J., *Fumifugium* (Oxford, 1661).

Gould, P. C., *Early Green Politics: Back to Nature, Back to Land and Socialism in Britain 1880–1900* (Brighton, 1988).

Hoskins, W. G., *The Making of the English Landscape* (London, 1955).

Leach, M. and Fairhead, J., *Misreading the African Landscape: Society and Ecology in the Forest-Savanna Mosaic* (Cambridge, 1996).

Meadows, D. H. and Meadows, D., *The Limits to Growth* (New York, 1972).

Merricks, L., 'Environmental History', *Rural History*, 3 (1996), 97–109.

Mowat, F., *Rescue the Earth: Conversations with Green Crusaders* (Toronto, 1990).

Ninan, M., *People of the Rainbow: A Nomadic Utopia* (Knoxville, TN, 1997).

Nora, P., *Les Lieux de mémoire* (Paris, 1992), translated by A. Goldhammer as *Realms of Memory* (New York, 1996).

Oelschlager, M., *The Idea of Wilderness: From Prehistory to the Age of Ecology* (New Haven, 1991).

Rival, L. (ed.), *The Social Life of Trees: An Anthropological Perspective on Tree Symbolism* (Oxford, 1998).

Santilli, M., *Àre* (São Paulo, 1987).

Schama, S., *Landscape and Memory* (London, 1995).

Sheail, J., *Rural Conservation in Inter-war Britain* (Oxford, 1981).

Thomas, K., *Man and the Natural World* (London, 1983).

Williams, M., 'The Relations of Environmental History and Historical Geography, *Journal of Historical Geography*, 20 (1994), 3–21.

1 The English, the trees, the wild and the green

Two millennia of mythological metamorphoses

Paul Thompson

Environmental activism is rightly seen as one of the most distinctive forms of political consciousness in the late-twentieth-century developed world. But as soon as one begins to ask how people conceive of the relationship between themselves and the natural world, and why they dedicate themselves to particular causes and campaigns, it soon becomes clear that the stories which they are telling are a remarkable mixture between the very old and the very new. Indeed even the very idea of urgency, that change has become especially fast and catastrophic in our own generation, goes back hundreds of years. Already in sixteenth-century England – which is about as early as one can find recorded oral traditions there – old men could be found bemoaning the loss of woodland in their lifetimes, and narrating mythical stories of travelling in boyhood from one village to the next through the woods without ever touching the ground.[1]

Contemporary environmental activism is both international and multifaceted. It ranges from militant direct activism to powerbroking with governments and international corporations; from the local to the global; from the scientific to the mystical and utopian. Its campaigns can focus anywhere from the metropolis to the wilderness and try to rescue anything from whales to butterflies. But whatever the campaign, environmentalism is able to rally support not only because of the immediate issue at stake, but because it draws on very deep – if only barely articulated – feelings about the relationship between human beings and nature, and hence about the meaning of human life itself. And obviously such feelings, conscious and unconscious, have been evolving for millennia. My purpose here is to trace through a single thread. It is the feelings – of both love and awe – which the English have long felt for trees. And more specifically, I want to trace the development of two of the most resonant images of woods and forests in England: the wild people and the green men.

It is important to be reminded that trees are not an inspiration in all cultures. The traditional image of the tree in a popular Czech proverb, for example, is as a symbol of the limitations and constrictions of life: 'trees don't grow to the heavens'.[2] But trees, woods and forests certainly have a singular power to move the English today. Over the last five years there has been repeated publicity in the national press for the defiant tree-dwelling young campaigners whose direct action resistance has become the bane of motorway-builders. For the more cautious and law-abiding, of all the charities in Britain the most popular is the umbrella conservation organization, the National Trust, whose two million members exceed those of the Church of England: its symbol, the oak leaf, not only conveying patriotic associations, but already at the time of its foundation in 1894 associated for folklorists with ancient British religious traditions. Younger but scarcely less impressive in terms of recent mobilization is the Woodland Trust. Founded in 1972, it now claims 200,000 subscribers and volunteers, holds 37,000 acres and takes an old wood a week into its care, as well as campaigning for the creation of completely new English woodlands. Equally striking among environmentalist activity abroad is the focus on forests – exemplified by Andréa Zhouri's account on the English who are working to save the Amazon rainforest. The NGOs (non-governmental organizations) involved range from the establishment Worldwide Fund for Nature to green organizations like Friends of the Earth, militant pressure groups like Greenpeace, and social NGOs such as Oxfam.[3] Among international conservation causes perhaps only the threatened extinction of the ocean whales mirrors in emotional intensity the fight to save the forests.

In rallying support for these causes scientific arguments have certainly played an important part. Indeed the whole contemporary debate is now set within a frame in which the fragility of the world environment and the potential for human activity to unbalance it irreversibly has become widely accepted, above all through the discovery of the breaking down of the ozone layer and the increasing threat of cancer which this brings. Although the focus has now shifted, at one time the role of the tropical rainforests in safeguarding the world's climate was highlighted, and this view reappears in lay discussions: it is said, for instance, that some 'scientists' have predicted that by the early twenty-first century, if the African, Indonesian and Amazonian rainforests are destroyed, 'the torrid zones of the earth will be reduced to desert or near desert and close to a quarter of the earth's population will be reduced to a state of destitution'.[4]

Scientific and pseudo-scientific overviews of this kind are usually relatively recent. In the past reason rarely favoured the trees. In fact the English have a history not only as tree-lovers, but equally as eager tree-fellers to meet their immediate economic needs. Even in the post-1945 period they were actively encouraging the felling of tropical forests in the interests of agricultural development, as in the notorious groundnuts scheme in East Africa: the need to protect the forests for world interests only became an active concern after the withdrawal from Empire. At the same time, although in the face of a mounting resistance by conservationists, at home English farmers bulldozed through the destruction of 150,000 miles of farm hedgerows throughout southern and eastern England, bringing serious fears of soil erosion as well as aesthetic devastation of the landscape.[5] Nor was this destructiveness new. Historically the English have been exceptionally ruthless in their felling. Even as early as the Norman Conquest, less than a fifth of England remained as woodland. By 1800 woodland was down to a mere 6 per cent, and a century later to a mere 4 per cent, the lowest proportion in Europe.[6] Thus contrary to what is commonly believed today, and was already believed in the early modern period, the major loss of English woodland took place many centuries ago.

Nevertheless, it is striking how some of the motifs of contemporary concerns with the forests turn out to have long antecedents. For instance, it has been argued recently that the Amazon needs to be protected because the forests harbour innumerable unidentified species and organisms, and within this biodiversity some of the forms may prove crucial to the future of humanity. But similar stories were current in the nineteenth century. In *The Way We Live Now*, Trollope's novel about City speculation, Henrietta Carbury meets a Dr Palmoil, who inspires her with a desire 'to visit the interior of Africa. It is the garden of the world . . . It seems that if we can only open the interior of Africa a little further, we can get everything that is wanted to complete the chemical combination necessary for feeding the human race.'[7]

Such fantasies seem to share elements with the search for the legendary King Solomon's mines, a story which goes back into the middle ages and which inspired several of the first European explorers of Africa, and also, encouraged by the unexpected riches seized by the first Spanish conquerors of Latin America, was then transposed to the New World.[8] This was a motif for Raleigh's disastrous expedition to the Orinoco in 1616, as well as a theme of English novels such as Rider Haggard's *King Solomon's Mines* and Conrad's *Nostromo*. Nevertheless the ancient search for gold and silver is not the same as the modern hope for the discovery of chemicals to transform human life. Precious metals are merely inert. In the contemporary imagination the forest may harbour secrets vital to human life itself, just as the entire forest has been imagined as a vast lung whose breath is essential to the working of the global climate.

Yet this apparently new meaning, the association of trees with life itself, is only a mutation on a theme which can be traced back – although as we shall see, with a good deal of imagination – to prehistoric times. Deciduous trees offer the most striking of all natural symbols of rebirth, and so of fertility, of the continuation of life beyond apparent death. In England, from the stone-carved capitals of medieval churches and cathedrals to the stained glass windows, wallpapers and chintzes of William Morris, leaf patterns seem perennially resurgent.

It seems to be John Aubrey, seventeenth-century chronicler and forerunner of oral history, who first maintained that the English love of oak trees was based on oral traditions going back to the Druids. For him the oak embodied magical powers and feelings: 'when an oake is felling, before it falls it gives a kind of shrieks or groanes, that be heard a mile off, as if it were the genius of the oak lamenting . . . To cut oak-wood is unfortunate.'[9] Certainly, despite the continuing hostility of the church to 'holy' trees – which in the middle ages zealous clergy had insisted should be cut down[10] – there is a striking continuity of oak trees invested with historic significance, including Gospel Oaks, the oak at Hatfield under which the young Elizabeth received the news that she was queen, oaks which gave shelter to the defeated Charles II, and even at Tolpuddle a Martyrs' Tree – which the TUC helps to protect – sacred to the pioneer farmworker trade unionists who were sent to exile in Australia in 1834.

The special sacredness of the yew is, by contrast, as an evergreen, alive even in the dead season. This has again been traced back to antiquity. It does indeed now seem clear from carbon testing that the yew trees in some churchyards are older than the churches themselves: sometimes over a thousand years old.[11] And sacred groves are as familiar to classical as to Celtic traditions. Indeed the hilltop site on the ridge of Monte Albano, overlooking Florence, which is now thought to have been the lost great temple of the Etruscans, is a wood of yew trees enclosed by a circle of embankments, rare in Italy but strikingly like the hilltop camps of the early British.[12]

One of the oldest images in Indo-European art and literature is that of the Tree of Life, 'a tree symbolic of life or immortality, *especially* that found in the Garden of Eden'. An outstanding early instance can be found in the frescoes of the tomb of Thoutmès III at Thebes, painted in about 1500 BC.[13] As a motif in art it was taken up especially early by the Persians. In India it became linked with legends of speaking trees, such as the oracular tree which warned Alexander the Great against his projected invasion.[14] The continuing power of this image in India was one strand in the extraordinary growth in the 1970s of the Gandhian Chipko movement, a populist environmental campaign to protect the Indian forests through passive direct resistance against the logging companies which took the form of mass 'tree hugging' by village women.[15]

There can, however, be many twists in the meaning of such symbolic trees. Thus in medieval Europe one variant was the depiction of Christ crucified on a live tree.[16] Here the tree becomes a symbol of redemption, giving everlasting spiritual life in exchange for life on earth. Perhaps there is a parallel here with the ancient myth of Apollo and Daphne, in which Daphne is seized by the lustful Apollo but saved at that very moment by sprouting roots and leaves.[17] Another was the genealogical Tree of Jesse, in which the tree springs from the lying figure of Jesse at its base and his descendants hang like fruit from the tree's boughs, which may be found in major churches throughout Europe.

The Tree of Jesse can be seen as a religious forerunner of the later secular 'family tree'. And from the time of the French Revolution onwards the traditional tree had won a new radical secular meaning as the Tree of Liberty. In twentieth-century England the Tree was taken up especially by Arts and Crafts architects[18] and jewellery and metalwork designers, many of whom were socialist followers of William Morris. There is a beautiful miniature oak candlesconce of brass designed in 1906 by Edward Gimson at Cheltenham Art Gallery, while a tree silver brooch by Voysey is now marketed as a commercial reproduction. The tree motif was also exported by the Spanish conquerors to Mexico, where it became associated with the particularly exuberant Mexican celebration of death on All Souls' Day, for which Mexican potters like those of Metepec make wonderful flowering tree wheels of life, full of birds and other creatures: gardens of Eden in themselves.[19]

However, the association of the tree with the Garden of Eden was complex. As the story of the Creation is recounted in Genesis:

> And out of the ground made the Lord God to grow every tree that is pleasant to the sight and good for food;
> the tree of life also in the midst of the garden,
> and the tree of knowledge of good and evil.[20]

The message was thus a contradictory one. Trees were a gift of God for human needs and pleasures. They were a symbol of eternal life and rebirth. Medieval sculptures of Adam often make the connection tangible, as he clutches the leaves on a tree bough to his genitals: often as if in fear, but sometimes even then in serene pride.[21] But equally, they symbolized the darker side of fertility, and fear of human sexuality, through the legend of Adam, Eve and the serpent, who hung twisted in the Tree of Knowledge, offering her one of the apples with which it was loaded: and so brought the fall of humanity.

This ambivalence seems to have been fundamental to earlier attitudes to trees and the forest. Sometimes indeed, in medieval manuscripts, like the twelfth-century *Liber floridus* from Saint-Omer, the good and evil trees are drawn in parallel, growing in opposite directions from the same intertwined root stock.[22] And ambivalence is in itself one explanation of why those mythical traditions which can be traced back from the present to the distant past have often undergone remarkable transformations during their transmission. These transformations are particularly interesting in the contrasting stories of the two symbolic anthropomorphic figures most strongly associated across the centuries in England with trees and the forest. They are, on the one hand the green man, and on the other, the wild man and the wild woman. Both go back as images at least to the middle ages. But while the green man has been a mysterious but essentially positive symbol, the qualities of the wild man and woman have been much more in contest.

Of the two, the green man is the more puzzling figure, for two reasons. The first is that his reappearance this century has changed his image and confused it with that held earlier by the wild man. A more fundamental difficulty, but a fascinating one, is that the green man tradition seems to have been transmitted in the middle ages almost entirely through visual images. They were transmitted, moreover, right across Western Europe. The masons who carved them left no words to help us understand them. But in their professional culture communicating with images rather than with words was scarcely remarkable. By the late middle ages leading stonemasons and woodcarvers could be people of considerable local importance, mayors of their cities; and a few of them certainly worked internationally. By then they were also taking images direct from printed books such as the *Biblia Pauperum*.[23] And even earlier, the gothic style of architecture had itself passed in its variant visual forms from country to country in Europe, although the masons who used it can rarely have shared a common vernacular language in their speaking.

We thus have in the medieval green man a mute legend, narrated through form rather than through words. There are only a few legends, such as the Arthurian medieval story of *Sir Gawaine and the Greene Knight*, or the modern Scottish tinker's tale of the Green Man of Knowledge, both about violence and courtship, which survive to suggest interpretations of the green man: and none is easy to connect with the green man image in sculpture. I know of no traditional children's stories about green men. Indeed there is no convincing evidence, even visual, that the green man was a meaningful symbol to ordinary people in the middle ages. In France and Belgium, for example, pilgrims commonly carried miniature lead pendants with a range of religious and secular symbols: along with hearts, doves, mermaids and sheela na gigs, these included wild men and women, but no green men.[24] Thus we can only interpret the image through what we can see in the sculpture of the masons.

The green man is represented in medieval sculpture almost always as male, and simply by his head: there is only one important full-length figure in England, the green man armed with a sword and buckler on the back of one of the choir stalls of Winchester cathedral, centre of medieval Arthurian legend, superbly carved among naturalistic foliage by William Lyngwode in 1308.[25] The green men heads typically take three forms: a face composed of leaves, with only the eyes peeping through; a head with vegetation growing out of its mouth, and sometimes eyes and ears too; or a head within vegetation, in the place of a flower. While the origin of the first of these forms is the late Romano-Greek Dionysian mask of vine leaves, found extensively in the Mediterranean from the second century onwards, the head disgorging leaves first appears in the early fifth century at Poitiers, close to the northern imperial frontier, on the tomb of St Arbre, daughter of the apostle of the Gauls. These two versions might therefore represent Christian incorporations of both pagan classical and Celtic beliefs. However, while the transmission of the image from classical sources is clearly established, the Celtic connection is much more speculative, for the foliate head does not appear in the interlaced foliage decorations of stone crosses and manuscripts in early Christian Irish or Scottish art.[26]

From the twelfth and thirteenth centuries until the end of the middle ages, however, green man heads became a favourite motif of sculptors, and they can be found on corbels, roof bosses and choir stalls in almost all the major ancient cathedrals and convent churches of France, Germany and England, and also in many minor churches too, as well as in some major buildings in northern Spain. In England green men are the most widespread single type of *all* the surviving images used by sculptors from the Norman period to the

Reformation, and they have been located in at least twenty-three counties. And they are wonderfully varied in form. In the west of England many are carved in wood, especially as roof bosses, but elsewhere they are mostly in stone. Some are quite crude and rough. Others are outstanding works of art: some probably portraits, as in the cloister of Norwich cathedral, others like the Kilpeck doorway conveying a robust sculptural power, yet others such as those of Ely or Southwell cathedrals embowered by the exquisitely delicate naturalistic leaves which surround them. Searching for green men is an aesthetic feast.

But what do they really mean? Different people often see different meanings in the same images. Kathleen Basford, for example, sees green man heads as typically baleful, 'full of dark foreboding': and she chooses photographs from unusual angles which make many of the faces look oddly twisted.[27] Nevertheless it is generally suggested that they are associated with fertility. William Anderson argues persuasively that this should be interpreted in the higher sense: 'the Green Man signifies irrepressible life . . . an image of renewal and rebirth . . . He knows and utters the secret laws of Nature', and visually he symbolizes 'the union of humanity and the vegetable world'. In one sense he may represent the survival of a pre-Christian pantheism, an alternative version of the vision of eternal life. But as Anderson also suggests, he may have meant something more specific to the medieval carvers: the green man 'was to them the symbol of the most precious gift they could receive from above, the gift of inspiration'.[28]

This is a speculative interpretation, but it makes sense, especially when one begins to look more closely at where the green men are found in medieval churches. They are most often in the eastern, that is the more spiritual parts of the church, as opposed to the more secular nave where the laity sat or stood; and especially in the choir, suggesting some possible affinity with the spiritual inspiration of music. Moreover there are also various indications which together suggest that they may well have had a special secret meaning for the creative skills of the masons themselves. First, they are typically half-hidden away behind their leaves, and most often in places where they are less easily seen: high up as roof bosses, or in side parts of the church, or on misericords underneath the choir stalls. Sometimes they seem deliberately positioned to see without being easily seen, as behind a screen or arch. Occasionally they were carved in places which only the sculptor could see, invisible to either priest or people.[29] Second, while green men can certainly be found in some minor parish churches, especially on the wooden wagon roofs of the west country, the great concentrations are to be found in the major cathedrals, priories and abbey churches. This was precisely where sufficient masons would be gathered for a common culture to develop with its own symbolism. And indeed, it is not by accident that one of the largest of gatherings of green men anywhere is at the Roslin Chapel, built outside Edinburgh from 1446 by William Sinclair, Earl of Orkney and grandmaster mason of Scotland. This exceptionally richly sculptured building is a *tour de force* of medieval sculpture which must have always been seen as such. One of its pillars is the Prentice Pillar, which according to local legend was carved brilliantly by an apprentice in his master's absence to show that he could work alone, but also fatally: for his master on his return murdered the apprentice from jealousy. The chapel also carries a ritual masonic inscription, and an extraordinary medley of biblical and masonic motifs: including at least 120 versions of the green man. At Roslin, exceptionally, the masonic connection was continuous from its medieval foundation to the freemasonry of the late nineteenth century.[30] It does seem reasonable to infer that for these medieval sculptors the green man symbolized their common skill and implied a higher form of life force fertility, artistic creativity.

There are, however, other possible interpretations, which we may see as alternative but not necessarily conflicting messages from the same image to other audiences. Two of the most powerful groups of green man sculptures are at the cathedrals of Southwell and Exeter. At the Southwell chapter house, built just before 1300, it is the wonderful naturalistic leaf carvings, mainly of maple, hawthorn and oak, which dominate: the green men seem by contrast puny and naive.[31] It may not be a coincidence that Southwell is the cathedral of the great Sherwood Forest, legendary haunt of Robin Hood, where medieval man must have felt more than usually smaller than nature. And why the predominance of maple, hawthorn and oak? The common maple, which no longer evokes interest, is a puzzle, but the hawthorn was certainly the prime symbol of spring, the may tree; while we have already encountered the oak.

Exeter is the city of Devon, where the parish churches have more green man carvings, and especially rather simple wood carvings probably by local artists, than any other part of England. The iconography of its cathedral, built in the same years as the Southwell chapter house, and spelt out in a series of abundant leaf corbels which march from the west end to the high altar, seems an extraordinarily open intertwining of Christian imagery with popular paganism. On a choir corbel the green man actually carries the Virgin Mary; and in the nave it is hard to see how the flowery king and queen could not have evoked the annual May spring celebrations – if only we could indeed be sure that these already took place in something like their later form. There are hawthorn and oak corbels here too in the choir; and more beautiful green men among the lady chapel roof bosses beyond.[32] And Exeter cathedral is not the only place in Devon where the green man seems identified with the May king. At the fourteenth-century collegiate church of Ottery St Mary the green man, uncrowned but with triple leaves from his mouth, nose and eyes, seems to look longingly across the lady chapel to his embowered partner-to-be. Others will undoubtedly decode these mute messages from the past differently: but to me they forcibly suggest that in the south-west, possibly because Celtic culture had proved more resilient, the medieval church chose to accept and incorporate rather than to attempt to stifle lay associations between the green man and popular spring fertility customs.

The medieval wild man was also associated with fertility, but much more usually in a darker and more animal way. Unlike the green man whose association was with trees or vines, he was a creature of the denser woods and the forest. At least up until the seventeenth century forests remained dark and sinister places in the European imagination, unproductive, the haunts of outlaws and highwaymen, frightening and melancholy.[33] The word 'savage' comes from the Latin 'silva', a wood. This is where the wild man or 'woodwose' lived.

The Oxford Dictionary describes the woodwose as a term going back to the eleventh century, 'a wild man of the woods; a savage; a satyr; a faun'. The term 'woodman' comes in as an alternative in the fourteenth century. Satyrs and fauns, however, are not synonyms, even though they shared the wild man's lusty sexuality: they are figures from Greek and Roman mythology, not especially associated with the woods, and only partly human, with goats' or horses' ears, tails, and sometimes legs; satyrs were also associated with bacchanalian drunkenness. Dryads were their female counterparts on a somewhat higher level: nymphs of semi-divine beauty who inhabited the woods and trees. But even in the middle ages there was some confusion between the wild man proper and the figures of classical legend. Thus Wycliffe chose to use 'woodwose' to describe Jerome's Latin 'pelosi', hairy woodland demons, in his first English translation of the Bible in the 1380s.[34]

Wild men are strikingly different from green men in many ways. In representations they are always full figures rather than faces, with hairy bodies, and carrying long wooden clubs or tree branches. There are wild women too, with long hair, and hair on their bodies except for their breasts. Their meaning is also much better documented. Unlike green men, they appear as often as females as males, not only in children's stories but also in legends, and in domestic and communal secular contexts as well as churches. Their geographical distribution is more limited, concentrated above all in the Alpine regions – where in earlier instances especially they were usually giants, so that an Alpine wild woman had such long breasts that she could throw them back over her shoulders – and to a lesser extent in central France, eastern England, the Pyrenees and northern Spain. Wild men and women also differ from green men in that – at least for the medieval period – they have been the subject of serious study by several scholars.[35]

Medieval wild men and women represent a specific strand in a long Indo-European mythical tradition, which can be traced back to Babylonian epics in Enkidu, the wild friend of Gilgamesh, who lived all his life with animals unaware of human beings. There are traces of this legend in the Old Testament figure of Esau, the twin brother of Jacob. In contrast to the smooth-skinned, successfully scheming and domesticated Jacob, who outwitted him to steal his inheritance, Esau was red-skinned, 'hairy all over like a hair-cloak', naively trusting, a hunter and 'a man of the open plains'.[36] Romulus and Remus, the founders of Rome, were children brought up by wild animals. Parallels with the wild men can be found in the Greek and Roman myths of semi-human satyrs and fauns. Another strand of legendary imagery came from the Hebrew idea of the desert as a place of suffering and punishment, inhabited by mythical goatlike beasts bringing temptations of bestiality, but also as a place of contemplation, penitence and salvation. Christian monks took to the desert from the second century, many deliberately living in extremely primitive conditions, and from the fourth century there are myths of wild anchorites in Egypt, hairy all over, which are echoed in late medieval manuscripts in Spain, Italy, England and Germany.[37] Images of John the Baptist also often show him as covered in long hair.[38] Yet another probable source in north-western Europe was the Celtic myths of the Druids as forest priests.

The hairy wild man of the middle ages first definitely appears in sculpture as late as the mid-thirteenth century, at Semur-en-Auxois in Provence, long after the green man. At Semur he is most likely to represent Esau, next to Jacob counting his coins. There are however earlier antecedents, in which semi-human wild men are presented among the hybrids of creation.[39] And the wild men certainly need to be understood in relation to these legendary figures.

Mythical traditions of the existence of 'monstrous races' in distant lands, 'marvels of the east', can be traced back to the earliest reports of India known in Europe in the fourth century BC. Besides hairy wild men and women, they included people with dogheads, with eyes in their bodies, with feet like sun-shades, pigmies, giants and cannibals, and many more. Many of these fabulous stories were borrowed from Indian epics. In successive centuries they were brought together in many collections, of which the most influential was Pliny's *Historia Naturalis* of the first century AD. These marvels and monsters were described in most medieval encyclopaedias and often depicted on medieval maps. In the later middle ages they also became the staple of travel stories, and there are beautiful illustrations of them in manuscripts recounting the journeys of Marco Polo and Sir John Mandeville. Nor were the myths of monstrous races undermined by the European explorations of the world in the fifteenth and sixteenth centuries: on the contrary, these stimulated more strange reports. In 1595 Raleigh himself conveyed news of a headless nation which had been sighted in the Amazon jungle. The monstrous races became the subject of serious treatises by physicians and scientists. And they were also a vein of oral tradition in which intellectuals took a special delight. Thus the fourteenth-century founder's regulations for New College, Oxford, specified how on dark nights

> when in winter, on the occasion of any holiday a fire is lit for the fellows in the great hall, the fellows and scholars may, after their dinner or supper, amuse themselves in a suitable manner with singing or reciting poetry, or with chronicles of different kingdoms and the wonders of the world.[40]

Nevertheless, although interest in the wild man grew along with that in hybrids and monstrous races, he was always a man, as the connections of his image with Esau or the desert anchorites suggest; and by the later middle ages there is never any question of his essential humanity in representation. His popularity as an image had grown dramatically by the fifteenth century. He can be found not only in churches, on fonts, choir stalls, memorials, capitals, gargoyles and flanking doorways, but also in house frescoes, chimneys, beams, tiles, candlesticks and drinking cups. In coastal Suffolk and Norfolk small wild men can be found as supporters of more than twenty baptismal fonts, and it seems possible that this unique concentration reflects a semi-pagan cult going back to the capture of a wild man at sea by the fishermen of Orford in the twelfth century, which was recorded by Ralph of Coggeshall. Their ritual position suggests that to churchmen they represented the unredeemed state of men before baptism.[41]

Visual representations of wild women also become common by the late fifteenth century. Both wild men and women appear as decoration and illustrations in medieval manuscripts, and also in early printed books. Lack of documentation makes it uncertain when the wild man appears in folk customs, but there is an exceptionally early instance as a woodwose mask in a yuletide drama of 1348 in England, at Otford in Kent. There is no presence of wild women in such public rituals.[42] In France they appear more often in poetry than visually. In oral tradition it seems likely that the wild man had become an established myth by the twelfth century, when the church was already condemning superstitious belief in him. The church remained actively hostile for centuries: as late as 1691 a young man was condemned to death in Sweden for fornication with a wild woman.[43] References to him can be found in the folk traditions above all of Alpine regions under many different names: as 'homo sylvaticus', 'uomo selvaggio', 'il silvano', 'il pantogan', 'il salvanel', 'ums selvadigs', 'dar sambinelo', 'il fanes', 'il gnero', 'sarvage', 'sarvanot', 'hombre salvaje', 'basa jaün', 'ancho', 'xaña', 'homme sauvage', 'skogsnufra', 'skougman', 'lesní muzore', 'wild man', or 'wilder mann'.[44] Legends of the wild woman are even more extensive, and can even be found in regions where there are no traditions of the wild man, such as Flanders or parts of northern Germany.[45] In short, in contrast to the semi-secret masonic cult of the green man, by the end of the middle ages and into the early modern era the wild man and woman had a strong hold on the imagination of people at all levels of society. But what did they signify to them?

Put simply, I would suggest that in the late medieval and early modern periods wild men and women were seen as representing the lost primitive side of humanity, the innocence of mankind before the fall, living free, at ease with nature and with their own sexual desires. Because they were uncivilized, and could openly express the primitive instincts which were in everyone, they were both feared and envied. As Enkidu to Gilgamesh, Esau to Jacob, the id to the ego.

Ambivalence, as we have seen of representations of the tree as a symbol, is a recurrent aspect of medieval art. This is especially true of depictions of sexuality, such as in dancing, where the contradictory pulls between pleasure and fear – of sin, of disapproval, of involvement and responsibility, of disease – were especially powerful.[46] Ambivalence is certainly very obvious when we examine representations of the wild man and woman. Quite often they are portrayed in an idyllic way as happy families living at ease with the wild animals of the forest. Thus a twelfth-century German epic places the wild man in a scene where, 'stretched out under a linden tree, lay a lion and a dragon, a bear and a boar, as pretty to see as could be'.[47] It calls to mind the old engraving of biblical paradise, with the lion lying down with the lamb, which hung on the staircase of my grandmother's house, and fascinated me as a child: 'It won't come in my time', she told me, 'but it might in yours.' A German engraving of 1470 portrays a wild man and woman and their children as an innocent happy family.[48] Elsewhere the wild woman is portrayed as the wild man's model gentle companion. They were admired for their ability to live off the roots, fruits and grasses of the forest, and for their closeness to animals and command of them: for example, they could ride deer with ease. One late medieval German wooden chest is carved with a jovial hunting scene in which two naked wild boys are shown with a beautiful young wild woman gracefully riding a unicorn sidesaddle. A fifteenth-century French tapestry shows wild men comfortably overpowering a dragon, a unicorn and a lion.[49]

Yet there was a threatening side to all this. They could equally be thought of as too close to animals: bestial. And their love for children led them to abduct human babies. In some regions they were especially feared as child-devourers, like the mythical Greek 'lamia', a kind of witch of the woods.[50] And this double face to desire was all the more apparent in the untamed sexuality of the wild people.

In poetry and myth, and also in visual scenes, wild men and women are seen above all as pleasure-loving and erotic. They could realize the dream and desire of free love-making, naked in the open air. Wild women have ample long hair, a symbol of sexuality, falling over their full breasts, and all of them are 'obsessed with a craving for the love of mortal men and go out of their way to obtain it'. In Grimms' tales such wild women descend from the German hills as long-haired seducers of respectable married farmers. In a Spanish story of 1496 the wild woman introduces herself, 'Call me by the name of desire.'[51]

However, this sexuality is also recognized as out of control and very dangerous. A German memorial brass to two bishops shows a group of wild men feasting and then carrying off a lady on horseback from under the eyes of an armed knight.[52] Wild men and women could both lure humans into ponds and drown them.[53] Wild men could impregnate human women as they slept, or plant changelings with them, or kidnap them. Some wild women were even worse. In the Franche Comté the 'dames vertes' could lure a man into the forest and, having consummated their love, drive them to death and haunt them. Others were revealed as deceivers, whose alluring beauty concealed the monstrous sagging breasts and crinkled flesh of old hags.[54]

These classic male nightmares of overpowering female sexuality have other parallels. Probably the best known are the mermaids, young women with fishtails, who appear widely in English literature and images from the fourteenth century. They are north European equivalents to the sirens of Greek and Roman legend, half bird and half woman, who would lure sailors to their doom by their enchanted singing. The mermaid was equally erotic; indeed in the sixteenth century the word itself was also used to mean a prostitute. Mermaids are typically portrayed as beautiful figures with long hair and naked breasts, often preening themselves with a comb and mirror.[55] In the oral traditions recorded by the Brothers Grimm and Hans Andersen, both mermen and mermaids appear always as seeking human love, luring their victims to their watery deaths.[56] The Irish 'sheela na gigs' are another instance.

The most spectacular of the classical myths of devouring female sexuality were those of the Amazons, and, like the sirens, these were sufficiently well known to have entered the English language by the fourteenth century. The Greeks believed the Amazons to live far to the north-east. They were a warrior people composed entirely of females, exceptionally tall, securing their own reproduction by seizing men, who, having satisfied the women's lusts, were then killed. When the children of these forced sexual unions were born, the infant boys were equally systematically eliminated. In both their behaviour and their giant size, one can see the Amazons as a dark Mediterranean sister of the northern wild women. For up to the end of the middle ages, however alluring she may have become, she also remained fundamentally dangerous.

Up to this point, then, the green men and the wild men and women would seem to represent distinctly different symbolic meanings. Indeed in the early years of the Reformation in Germany, Protestant devotional and propaganda works associate Luther with the green man and portray the pope as a demoniacal wild man with a long tail.[57] Wild men also appear on two of the grandest public halls of the later middle ages – in Lübeck and Venice – probably as symbols of human origins and evolution in contrast to present achievement.[58] But such clear dualities were not to last. Subsequently, these meanings evolve in a much more complex way – which has been less clearly traced by scholars – under a variety of forces for change: through the Renaissance, the long phase of disapproval of popular customs, the transformation of attitudes of the educated to nature, the global expansion of Europe, and finally the rise of environmental movements.

The Renaissance transformed the visual vocabulary of European art by suppressing many decorative motifs and reintroducing others of pre-gothic classical derivation. It found no place for the wild man and woman, who are not found in high art after the sixteenth century. Dürer and Breughel were the last major artists to portray him. Albrecht Altdorfer, one of the pioneers of European landscape painting, dutifully transformed the wild families in his German forests into families of satyrs.[59] The wild man does continue to appear as a heraldic supporter for the arms of some families who adopted him while fashionable. This heraldic motif was especially popular in northern Spain, as in the fifteenth-century monuments to nobles and clergy at Avila cathedral, whose great west door is also flanked by giant wild men: but heraldic wild men can be found in fifteenth-century England too, for example in a Devon bench end at Colebrooke, or as giants flanking the gatehouse to Hinchingbrooke outside Huntingdon, and they spread elsewhere in England and Scotland in the next two centuries. The use of family surnames such as Woodiwiss even up to the present represents another form of continuity. And he survives more strongly in literature, where he can be found in both Spenser's *Faerie Queen* and in Cervantes, building the castle of love in *Don Quixote*. He resurges in nineteenth-century romantic literature, most strikingly in the prose romances of William Morris: of which more later.

The image of the green man, on the other hand, was confused by the Renaissance through the revival of often very similar bacchanalian leaf masks as decoration. In ancient Greek and Roman sculpture these masks had represented gods of alcoholic and erotic pleasure. They probably carried the same meanings in central Italy, where there had been no representations of green men in the middle ages, when they reappeared in fifteenth-century Florentine sculpture and painting: for example, in frescoes by Mantegna, or in the Medici tombs by Michelangelo. Unlike Anderson, I do not think that these can be safely regarded as green men.[60] To my eye, they feel explicitly classical and also, quite unlike the medieval green men, mechanically repetitive. The same doubts must apply to similar motifs introduced by leading architects to northern Europe in the sixteenth and seventeenth centuries.

On the other hand, it is important to recognize that part of this new mechanical feeling was a question of change not so much in style as in social control: for it was in this period that the modern concept of a single architect in close charge of all the detailing of a major building became accepted. It can be seen as a sign of hardening class distinctions. And it does seem very possible that in artisan work, domestic as well as ecclesiastical, where the change in style was more gradual and less intellectually driven, the old meanings of the green man could have continued in the new forms. This is especially clear in west country bench ends, tombs and rood screen ends with their medley of early sixteenth-century gothic and renaissance details, such as the screen at Atherington, the benches at Crowcombe and High Bickington, and a tomb of 1564 in Exeter Cathedral. Faces disgorging leaves can also be found in house panelling into the seventeenth century. The motif also appears in the keystones of artisan-built town housing, as at Queen Anne's Gate and Lincoln's Inn of the early eighteenth century in London.[61]

From the beginning of the eighteenth century, however, authentic green man foliate heads disappear from the visual arts. Apart from two maverick instances in classical gardens – one is the cast iron gates of Kew of 1843 – green men effectively disappear from the decorative repertoire until the end of the nineteenth century. One would have expected them to have been revived by the architects and designers of the Gothic Revival, but this was surprisingly rarely so. The green man could have fascinated William Burges, who lived in the imaginative world of the medieval bestiary, but there is scarcely a hint of him in Burges' designs.[62] His absence is equally striking in the designs of William Morris, whose stained glass, chintz and wallpapers are abundant with foliage, and sometimes forest creatures: but never a green man. Some Victorian clergy might have deliberately avoided foliate heads as a motif had they been aware of the green man's pagan origins, but certainly not Morris. Equally strikingly, when the brothers O'Shea came to carve foliage and animals on arches and capitals through the new University Museum in Oxford in the 1850s, they too included no green men. The O'Sheas, personally encouraged by John Ruskin as contemporary examples of the spirit of free craftsmanship, would have also delighted in a figure with such provocative associations: at Oxford they were finally sacked for carving academic parrots in mortar boards, and they went on to carve a monkey playing billiards outside the premier Dublin club. It thus seems that at all social levels the medieval green man of creative inspiration had disappeared from consciousness.

Possibly there was some renewed awareness of him among visual artists from the 1890s. Certainly foliage and leaves become more often mixed with figures and faces. I have in my possession a strange ceramic head, whose pale face is encoiled in green hair strongly suggesting foliage. But the Tree of Life was a much more favoured arts and crafts motif, and although one can occasionally find foliate heads from this period which do not have the goat's horns of satyrs, there are very few definite instances of green men.[63] We have to wait for this until the early-twentieth-century figure carved by William Simmons at Cecil Sharp house: and by this point the image has changed completely. The green man here is no longer a foliate head, but instead a jovial folkdancer covered in leaves. And an explicit recognition of the older visual tradition of the green man only began with Lady Raglan's article, published as late as 1939, on 'The "Green Man" in Church Architecture'. She had been shown a foliate head by the vicar of Llangym church on the Welsh border, himself a folklorist, 'who suggested that it was intended to symbolize the spirit of inspiration'. But Lady Raglan disagreed: 'it seemed to me certain that it was a man and not a spirit, and that moreover it was a "Green Man". And so I named it.' She went on to demonstrate the evidence she had gathered through folk custom.[64] But by this point we have reached another stage: the reinvention of the green man tradition.

This evidence is in fact much more confusing than Raglan believed, for she cheerfully amalgamated 'the figure variously known as the Green Man, Jack-in-the-Green, Robin Hood, the King of May, and the Garland, who is the central figure in the May-day celebrations throughout Northern and Central Europe'.[65] While these traditions do all intersect, they each took on special meanings which are lost in such a fusion. This is obvious in the case of Robin

Hood, who was a forest wild man administering rough justice for the benefit of the poor rather than a fertility figure – and provided the first association of the woodland green with liberty. Moreover she omits to include the wild men in her list. They were certainly also part of the broad tradition.

It is certainly possible that some forms of fertility dances in England go back to pre-Christian times. There is a seventh-century cross shaft at Codford St Peter, Wiltshire, carved with a dancing figure, head stretched upward, holding up a leafy branch: at its top are buds and bells, and in his other hand he holds an instrument. This is probably the earliest-known representation of the spring dances. As early as 1240 the reforming Bishop Grosseteste of Lincoln protested against those of his clergy who joined 'games which they call the bringing-in of May'.[66] But clear accounts of these rituals first occur, usually from hostile observers who attacked their loudness, roughness and sexual licentiousness, only in the fifteenth and sixteenth centuries. There are also descriptions in Chaucer and in Malory's *Morte d'Arthur*. In Switzerland, southern Germany and Austria they relate to Carnival at the end of February, in England to Mayday, but the details are similar: the use of leaf garlands and cones, a May Queen and a venerated Maypole (only in England: where since Freud it has been widely presumed – although without evidence – to represent a phallus), and dancers with bells dressed in green moss and leaves.

Dancers in green moss and leaves are also described appearing at important civic occasions: at the Basel church council of 1435 (where they had long green hair), at Toledo in 1545, at a papal festival in Viterbo in 1462 and at a Sforza marriage in Pesaro in 1475; in England at the Twelfth Night celebrations before Henry VIII in 1515, at a masque before Queen Elizabeth at Kenilworth in 1575, at the Lord Mayor's Show in London, and at many other English town May ceremonies from the fifteenth century until a St George's Day civic procession in Chester in 1610, and as late as the 1680s in London. Like wild men, they can also be found as heraldic supporters in the seventeenth century. And in nearly all these instances the dancers are described, not as green men, but as 'wild men' or 'salvajes', and usually carry 'great clubs'. This is moreover the form of carnival dance which still continues in Alpine Europe. Exceptionally, however, in the last two occasions in England the contemporary commentators describe the figures, whose purpose is 'to maintaine way for the rest of the show', as 'Savages or Green Men'. By the end of the seventeenth century this confusion seems to have become usual: for example John Bagford, writing on the same figures as used in shop signs, says: 'they are called woudmen, or wildmen, thou' at this day we in the signe [trade] call them Green Men'.[67]

Whatever the symbolism of the May dances may have meant, there can be no doubt of their continued association for the young with lovemaking. Robert Herrick put it nicely in his 'Corinna goes a Maying' of the 1630s:

> Many a green-gown has been given;
> Many a kisse, both odd and even:

Many a glance too has been sent
From out the eye, Love's Firmament:
Many a jest told of the Keyes betraying
This night, and Locks pick't, yet we're not a Maying . . .

In England, however, partly because increasing class division led to a distaste for much popular custom and a new preference for classical masques, and partly because of attacks by Puritan moralists, the old dances disappeared from the view of educated commentators, and are rarely described again before the late eighteenth century. By this point a new romantic interest in popular traditions as survivals from an earlier culture led both to the collecting of oral poetry and ballads and to the description of customary rituals. Joseph Strutt's encyclopaedic *Sports and Pastimes of the People of England* of 1801 describes the May celebrations in great detail. He refers back especially to seventeenth-century sources. Two points are particularly interesting. First, he does illustrate both a wild man with his club and a 'green man', a dancer wearing a wreath of leaves on his head: but both are now described as actors in firework displays. On the origins of the May customs he comments: 'This custom is a relick of one more ancient, practised by the Heathens, who observed the last four days in April, and the first of May, in honour of the goddess Flora . . . Some consider the may-pole as a relique of Druidism; but I cannot find any solid foundation for such an opinion.'[68]

Thus the antiquarian and folklorist assumptions of a continuing tradition of May celebrations from the Druids to the present has itself a long pedigree. But it received a double stimulus in the 1890s. Sir James Frazer's classic study of cross-cultural universal customs in magic and religion, which is full of accounts of tree worship and appropriately entitled *The Golden Bough*, was first published in 1890. His comments on Jack-in-the-Green as the symbol of the sacred tree, death and resurrection, became the established view for the next forty years. And then Cecil Sharp accidentally came across a traditional Morris side still dancing in the street in Headington, Oxford, on Boxing Day 1899. The dance and its revival became emblematic of the English folklore movement. By this point, while other elements in the dance had continued, the dancers called wild men had disappeared: instead there was a new central figure, sometimes but often not dressed in leaves, and known as Jack-in-the-Green, who danced inside a mobile garland called 'the Bush'. Douglas Kennedy, the leading spokesman of the English Folk Dance and Song Society founded in 1912, believed that 'the green bush is itself the token of the reawakened life, symbolised throughout the Northern Temperate Zone by green shoots, springing buds and fruit-blossoms'. The Bush was also used as the sign for the alebower used to refresh the dancers, and the barman was called 'Mr Green' or 'Jack Green'.[69]

It now seems most probable that this dance was an evolution of the earlier – and especially late medieval – spring dances, probably during the eighteenth century, and especially in the southern English towns. There was a strong association between the London chimneysweeps and these dances, no doubt because of the sexual connotations of thrusting brooms up chimneys.[70] The spread of Green Man pubs – there are more than thirty in London alone – may also reflect the popularity of the dances, although they more likely originated from distillers' and herbalists' signs, and as pub signs they most often portray Robin Hood.[71]

The effect of the folklore revival was therefore to fuse the earlier images of the green man and the wild man, of creative intelligence and free sexuality, and to attribute both aspects to a reinvented 'popular' tradition of the green man.

In the meantime, however, the context of the woodland figures was itself being seen in very different ways. Between 1500 and 1800 there was a transformation in European attitudes to nature. 'New sensibilities arose towards animals, plants and landscape. The relationship of man to other species was redefined; and his right to exploit those species for his own advantage was sharply challenged.' In the middle ages the forest was seen essentially as a waste, a wilderness, the habitat of raw nature, a place of danger and fear. From the fifteenth century in the visual arts landscape painting develops, and the forest begins to be shown, for instance in tapestries, as a place of pleasure as well as hunting. In southern England the felling of trees as fuel for the growing metal industry was already observed with royal alarm: as James I put it in 1610, 'If woods be suffered to be felled, as daily they are, there will be none left.' Although instances of laying out deer parks and planting trees can be traced back into the middle ages, it was in the late seventeenth century that systematic treeplanting became an upper-class fashion, especially encouraged by John Evelyn's *Sylva*, a widely read book commissioned by the Royal Society and the Navy, which he claimed himself had inspired the planting of 'millions of timber trees' on both crown and private landowners' estates.[72] Subsequently it was English landowners who, in the eighteenth century, led the fashion for the picturesque, the design of parklands with a deliberately 'natural' irregularity, and by the later eighteenth century tree-drawing had become a strong fashion. It was a special skill of the English watercolourists; perhaps its greatest exponent was John Ruskin.

Meanwhile increasingly feelings began to be attributed to both animals and trees, and human attachments to them. Montaigne argued as early as the 1590s that trees and plants should be treated with humanity, and several seventeenth-century writers asserted that they were 'fellow creatures' capable of both affection and suffering. Alexander Pope declared a tree to be 'a nobler object than a prince in his coronation clothes', while an essayist of 1787 pitied a man who could not fall in love with a tree. From the Elizabethan Michael Drayton – who regretted when 'this whole country's face was forestry' – onwards, a swelling poetic chorus protested against the cutting down of noble trees: Cowper, Wordsworth, Clare, and perhaps most poignantly of all the Victorian priest Gerard Manley Hopkins:

My aspens dear, whose airy cages quelled,
Quelled or quenched in leaves the leaping sun,
All felled, are all felled, felled;
Of a fresh and following folded rank
Not spare, not one . . .
Oh if we but knew what we do
When we delve or hew –
Hack and rack the growing green . . .[73]

So from being places of fear, woodlands had become sources of human inspiration. And this was equally true in North America, where Ralph Waldo Emerson wrote, 'In the woods we return to reason and faith.' The pioneer preservationist John Muir called the Californian forests a 'Godful wilderness . . . God's first temples'. Henry Thoreau's *Walden* is a deeply reflective account of his experiment in living wild in the woods for two years in the 1840s. 'If we do not go to church as much as did our fathers', another New Englander observed in 1912, 'we go to the woods much more.'[74] For intellectual Americans the woods offered solitude and spiritual regeneration, much as the wildness of the desert had to the early Christian anchorites.

Thus it was not only the woods but also woodlanders and woodland living which was now romanticized. The most powerful influence in this change was that of Rousseau and his vindication of 'l'homme sauvage', to which we shall return. But we can also find the idea with nineteenth-century novelists, as when Thomas Hardy entitled his lyrical early tale of peasant love *Under the Greenwood Tree* (1872). In the 1880s and 1890s William Morris used his prose romances, his semi-poetic stories of the peoples living by the wild-wood, to elaborate some of his vision of communal societies.

It is clear from the romances that Morris well understood the double meaning that the forests had held for medieval people, and indeed he knew many of the traditional legends through Grimm, whose collection he described as a 'bible', and presented to his daughters. But he relegated the most sinister side of the forest to 'wood-wights', bewitching spirits like those described by Stoneface in *The Roots of the Mountains*:

> I see thou longest for the wood and the innermost of it . . . Such things are in the wood, yea, and before ye come to its innermost, as may well try the stoutest heart . . . There are moreover the lairs of Wights in the shape of women, that draw a young man's heart out of his body, and fill up the empty space with desire never to be satisfied, and they may mock him therewith and waste his Manhood and destroy him.

Sometimes tauntingly the wood-wight will reappear as a 'foul old hag' to the wanderer with his 'empty heart and a burning never-satisfied desire'. More typically, however, Morris gives magical powers to the 'wood-women' in the prose romances to enable them to act as strongly as the men, and they nearly all use these magical powers benevolently. Almost all of them are expressively erotic, kissing and caressing their men freely, and using their magic only to hold them: as the fair-haired Wood-Sun in *The House of the Wolfings*, and Habundia, the 'wood-mother' in *The Water of the Wondrous Isles*. Most attractive of all are the beautiful, active young women heroines, like the Maid in *The Wood beyond the World*, and especially Birdalone in *The Water*: who loves to swim naked in the forest pools and picnic from forest berries, first discovered in the wild-wood unclothed but for an oak-wreath around her sun-browned loins. The long tale follows the romance between her, the Lady of the Woods, and Arthur, whom she in turn rediscovers by a forest stream, playing his harp, clad in deerskins, with his brown hair 'hung down long and shaggy over his face'.[75] For Morris, the wood and the woodlanders have become the realization of communal living, sexual equality and ideal free romantic love. Malevolent spirits remain, but their power has softened: they never kill. These woodlanders certainly carry traits of the earlier wild women and men; but there is never a sign of a green man among them.

As a socialist reformer Morris advocated woodland camping for children as an essential part of education. Some of his admirers played key parts in the development of the Woodcraft Folk, the early-twentieth-century Anglo-American youth movement which since the 1930s has become the socialist alternative to the hierarchical and semi-militaristic Scouts in Britain. Its original philosophy intertwined romantic medievalism, including an Order of Woodcraft Chivalry, with the belief that, in order to become fully mature, every boy or girl should, following human evolution itself, spend a period learning to live in the woods. One of the founders, Ernest Thompson Seton, published numerous stories of woodland animals and Indians, including an especially popular tale of two white boys brought up by them in the woods.[76] Closely similar ideas also inspired the Forest School camps.[77] This belief in the social value of woodland life has become closely associated with contemporary new imagery of the green men.

Even in eighteenth-century England, reported sightings of wild men and women had not wholly ceased. They survived in the popular imagination in two ways. First, the traditional medieval tale of Valentine and his brother Orson, lost as an infant in the forest and brought up by a bear, was being passed on through popular pamphlets: one of 1804, entitled *The Famous History of Valentine and Orson, or the Wild Man of Orleans*, and published along with sea stories, and books of recipes, songs and jokes by T. Hughes of London, has a frontispiece of a mounted knight in armour confronting a hairy wild man with claws. Similarly in Birmingham in 1812, Biddle and Hudson included the tale in their *Juvenile Books* series, along with *Rural Walks in Spring with Moral Reflections*.[78]

In parallel with these old tales, popular pamphlets – often with wonderful titles, such as *It Cannot Rain but It Pours: Or, London Strow'd with Rarities*, in which the arrival of 'Peter the Savage' is discussed alongside the arrival of a rare white bear, an Italian woman singer, and a 'copper-farting dean from Ireland' – described discoveries of wild human children who had been brought up in the woods by animals. Usually the link with the tale of Orson was made explicit. Two of the best known were the wild boy Peter, who was found in Germany, brought for exhibit and investigation to the royal court of George II in 1726, and seduced a dairymaid in Harrow; and *La Belle Sauvage*, a 9-year-old girl found in the woods of Champagne and investigated by the French academy: both still the subject of popular pamphlets in the nineteenth century.[79] These findings, however, excited interest not only at a popular level and at the court, but also among scientists. Another French wild boy, found in Aveyron in 1799, was subjected with very partial success to a sustained educational programme by Jean-Marc-Gaspard Itard, physician of the leading Paris hospital for deaf-mutes: an experiment which remained of scientific interest into the twentieth century.[80] And behind Itard's reflective seriousness we can almost certainly trace a major new influence on thinking about wild men and women, that of Rousseau.

Rousseau's thoughts on the 'noble savage', *le bon sauvage*, were a contribution to a long literary debate, which literary scholars have traced through many different forms: accounts by travellers continued to be one form, but in both English and Spanish poetry and the theatre were also very important.[81] His most crucial innovation was not merely to see, as many others had done, the positive aspects of life in the wild, but to suggest that there is a wild man or woman within every civilized feeling person; and that it is a strength in us. It was above all through Rousseau that the wild people of the forest came to be seen not so much as dark savages, but rather as innocents incorporating all the best potential of humanity. From Rousseau onwards we can say that in one sense there was a personal awareness of the wild in the heart of European intellectual consciousness. In this he marks a crucial step on the path to Freud's closely comparable analysis of *Civilisation and its Discontents*.

Paradoxically at precisely this moment Europeans had come to believe that wild men and women were most likely to be found, not in their own forests, but much further away. They had always been seen on the periphery of the cultivated and civilized world. But as the frontier of European knowledge pushed outwards, the wild men retreated in the imagination. Originally the direction was to the east, where the Amazons were also believed to live. We have seen that the belief that strange races could be found in India and North Africa was widespread, supported by Pliny and also by Ptolemy, whose *Monstrous Races of the East* was printed in Nuremberg in 1494. One English traveller, Edward Webbe, even claimed in 1590 that he had found the court of Prester John and seen a wild man there chained to a post. Encouraged also by Breidenbach's *Journey to the Holy Land* of 1486, an over-imaginative racial science developed, creating a hierarchy of races which ranged from apes and monkeys, through the varieties of 'homo monstruosus', including satyrs, troglodytes, pigmies and wild men, to the Irish, and finally the English,

Germans or Italians. One example by the Jesuit Gaspar Schott, *Homines Sylvestres ac Pilosi* (1667), illustrates a wild woman, while another by Edward Tyson, *Anatomy of a Pygmie-Orang Aoutang, Sive Homo Sylvestris* (London, 1699), purports to show a pigmy wild man from Java.[82] It was into these overwhelmingly derogatory categorizations that Europeans began to fit their evaluations of the new peoples whom they were now encountering in other continents: and above all, in Africa and the Americas. On the staircase of Salamanca University there is a wild man standing guard; but instead of a formidable wild man holding a club, he now holds the net, bow and arrows of an American Indian; and he looks apprehensive. To the first settlers of New England at Plymouth Colony there was nothing to be approved in either the native landscape or its people: a 'hideous and desolate wilderness . . . full of wild beastes and wild men'.[83] Such assumptions helped to justify the genocide of the American Indians which followed.

Not only attitudes, but in the case of sixteenth-century New Spain European images of the wild man were also exported to the colonies. Mexican Indian craftsmen were put to carving grand figures of hairy men outside colonial palaces, and at the great spectacle organized by the Viceroy in Mexico City in 1538 to celebrate peace with France, the city's central square was filled with a forest of trees in which were deer, lions, foxes, tigers and rabbits, and 'a squadron of wild men with knotted and twisted clubs'.[84]

From the start too there was a more sympathetic countercurrent, picking up precisely those features of the medieval wild man which had been felt most attractive. Thus Montaigne's *Des Cannibales* of the 1590s gave a positive account of the primitive communal economy, which was quoted by Shakespeare in his rendering of the wild man Caliban in *The Tempest*.[85] In Spain these themes were especially taken up in the plays of Lope de Vega, whose cave-dwelling Caribs are clearly colonial wild men.[86] Before long the Jesuits were attempting to create a new communal paradise for the Indians in Paraguay. The first English sculptures of Indians from the Amazon, on a merchant's tomb at Burford of 1569, convey an almost sweet docility.[87] And then there was the tale of Pocahontas.

The sexuality of the wild men and women had always been as much envied as feared: and these same ambivalences were projected on to the American Indians. The earliest European representations of native Americans in woodcuts of the 1490s showed Columbus' boat landing while naked figures danced ashore. Vespucci reported that their women were unashamed, naked and 'extremely libidinous'. Other travellers emphasized the prevalence of cannibalism; and early woodcuts projected both images in parallel.[88] Such travellers' reports proved a long-standing tradition, whose discourse was later absorbed into the imaginative framework of western anthropology of 'the other'.[89] In the same spirit, the great river system of the southern continent was named in 1541 when Orellana sailed from its headwaters in Peru to its mouth at the Atlantic. During his long and dangerous voyage, probably delirious, he believed that he had been attacked by tall naked women warriors;

and so he transplanted the Greek legend, to name the river the Amazon. Such female warriors have never been subsequently found among the forest peoples; but nevertheless, the myth of the sexual power of Native American women was to survive.

It was not long afterwards, in 1607, that an English sailor, John Rolfe, fell in love with a North American Indian, the wild princess Pocahontas. She was her father's 'delight and darling', and even before this famous among the English colonial adventurers as '*Nonparella* of Virginia'. Rolfe had been captured by her father's warriors, and his life had been threatened, but he was freed at her behest. His letter pleading the colony's governor to let him marry her makes his physical passion explicit. He describes himself as sleepless from 'the unbridled desire of carnal affection' for Pocahontas, 'to whom my hartie and best thoughts are, and have a long time bin so intangled, and inthraled in so intricate a laborinth, that I was even awearied to unwinde myself thereout'.[90] Rolfe was a sea captain from Gravesend at the mouth of the Thames. Here my father was brought up close to the ships and the river, and he would recount the story's sequel to me, of how once brought to England, clothed and baptized, Pocahontas lost her wild vitality, pined away and died. This same interpretation of the story – the stifling of sexuality through Christianization – is represented in the memorial windows in Gravesend parish church where she is buried, which were presented in the 1930s from America.

Walt Disney's recent cartoon film *Pocahontas* significantly transforms the scenery from New England to a tropical forest and gives the romance a happy ending. And indeed, more recently Latin America, and especially Amazonia, has become increasingly the focus for the search for a lost human innocence in sexuality and in oneness with nature. But this has been the culmination of a slow and meandering process.

It is clear that by the nineteenth century – if not much earlier – the Indians, after centuries of warfare and effective extermination over most of the continent, had lost their romantic magic in the minds of most North American settlers. By this point there were three regional variations of the dominant representations of trees and wild people. In the west, where significant numbers of Indians did survive, they were portrayed in paintings and stories as warrior people, and usually well clothed;[91] but by now – paving the way for the twentieth-century genre of wild west films and thriller novels – it was not the Indian but the valiant lone pioneer white frontiersman who was, to cite Robert Ballantyne's classic of 1863, the hero of *The Wild Man of the West: A Tale of the Rocky Mountains*. Such wild white heroes run on seamlessly towards a recent title like *The Wild Man at Smoke Creek*, and can boast Australian offspring too.[92]

In the Old South, where the Indians had been swept out to make way for black slave plantations, there was no public romanticization of either black or Indian, and the tree itself became a most sinister symbol, a dark reversal of the tree of the cross on which Christ was crucified, as the lynching tree on which bold blacks were strung up dead. As Billie Holiday sang, 'Southern trees bear bitter fruit'.[93] Maybe, given the powerful hand of capitalism behind the slavery of the plantation system, it is no accident that we can also find in the South today the ultimate marketization of the family tree. A leaflet issued by the Kentucky History Center calls for support for its new building in Frankfort, which will include a symbolic brass tree, and every supporter is 'invited to purchase a personalised leaf' for a minimum of $200, for which they will receive 'an official Certificate of Ownership'.[94]

Only in New England did romanticization of the American forest flourish, as epitomized in Thoreau's *Walden*, or in the landscape painting of the popular artist Frederick Church.[95] But even here we find no Indians in either book or painting. Thoreau was in fact surprisingly uninterested in trees as such, but much more in the feasibility of living the simple life: his only forest people are old whites, storytellers who pass on to him forest lore and oral traditions.[96] Other New Englanders did see the woods as intrinsically mystical and religious, cathedral-like: for example Washington Irving, or the Ruskinian early-twentieth-century painter Charles Burchfield. For Burchfield as a child the fortitude of an oak leaf in resisting the fall winds had given him courage to fight severe illness, and in adulthood he believed that contemplating a tree was 'more of a prayer than meaningless phrases mumbled in a church'.[97] But there are no wild people in his cathedral glades.

In Latin America, by contrast, the Spanish and Portuguese settlers had not exterminated the native people of the forests: and it was therefore to the tropical Indians that European attention gradually shifted. Opinion continued to be sharply divided between those who thought that they lived like brutes, and those who viewed them as living in prelapsarian innocence, mankind as before the Fall: as the saintly John Donne put it in 1598:

> So naked to this day, as though man there
> From paradise so great a distance were,
> As yet the newes could not arrived bee,
> Of *Adam's* tasting the forbidden tree.

The same split continued in the later seventeenth century, when Hobbes' *Leviathan* portrayed 'the savage people in many places of America' as still living 'at this day in that brutish manner', while Dryden first coined the term, 'the noble savage':

> I am as free as nature first made man,
> Ere the base laws of servitude began
> When wild in woods the noble savage ran.[98]

Dryden's was an image to echo, not only for Rousseau, but on to the English romantic poets of the nineteenth century, as when Wordsworth dreamt of a childhood 'on Indian plains, and . . . run abroad in wantonness . . . A naked savage, in the thunder shower'.[99]

In the late eighteenth century, however, a major diversion had intervened, through the discovery of a rival tropical paradise in the South Seas. With the publication of *Hawkesworth's Voyages* in 1773 the Polynesian islanders became the pre-eminent noble savages. While Captain Cook had been distressed by the sexual licence of their women and their provocative dancing, Hawkesworth was much more sympathetic, arguing that they did have a definite 'knowledge of right and wrong', and that 'if we admit that they are upon the whole happier than we, we must admit that the child is happier than the man, and that we are the losers by the perfection of our nature, the increase of our knowledge, and the enlargement of our views'. The idyllic paintings of William Hodges, who went on Cook's second and third voyages, portraying naked girls bathing against a Tahitian background of palms and volcanoes, equally 'engrossed conversation from the politicist circles and throughout every class of the kingdom'.[100] The fascination with South Seas wild people continued through the nineteenth century, stimulated by accounts of feral children, critical attacks by missionaries, and highly sexualized travel accounts like James Greenwood's *The Wild Man at Home: Pictures of Life in Savage Lands* of 1879.[101] It was no accident that it was at Tahiti that Gauguin arrived in 1891 to paint noble savage women. Derain followed him too, at least in the imagination.[102] Nor that when Margaret Mead wanted an anthropological testing-ground for a freer North American adolescence, she sought and found it in the Pacific islands: the wild people myth thus becoming classic anthropology.[103]

The lure of the South Seas was already fading well before that tropical paradise was polluted as another western testing-ground, this time for nuclear weapons. The noble savages of the Amazon had in fact never been entirely forgotten by Europeans. There was a continuing series of operas – including Mozart's unfinished *Il Regno delle Amazoni* – set in Amazonia by Italian, French and German composers, from Cavalli's *Vernanda l'Amazzone* of 1652 through to the late nineteenth century. The roll of painters taking Amazonian themes similarly runs from Rubens to the heroic classicism of Colin, Géricault, the sensual Victorian William Etty, the Arthurian Burne-Jones, and even to the twentieth-century vorticist Wyndham Lewis and surrealist de Chirico.[104] The English naturalist W. H. Hudson's popular 'romance of the tropical forest', *Green Mansions*, about a tree spirit who appears as 'a wild, solitary girl of the woods' but only speaks in birdsong, was set here too. And meanwhile the old idyll also continued to be portrayed in travellers' accounts rediscovering the Amazon, such as by Alfred Russell Wallace in the 1850s, or the happy naked wild families drawn by Debret and Rugendas, or the word-painting of Theodore Roosevelt's *Through the Brazilian Wilderness* of 1914, a vision of 'Adam and Eve before the Fall':

> It was an interesting sight . . . to see these utterly wild, friendly savages, circling in their slow dance, and chanting their immemorial melodies, in

> the brilliant tropical moonlight, with the river rushing by in the background, through the lonely heart of the wilderness.[105]

From the 1960s, however, the pace of these rediscoveries of lost Edens on the Amazon has greatly quickened. Fiction includes *The Wild Man* by the Mills and Boon novelist Margaret Rome, Mario Vargas Llosa's brilliant reflections on an almost extinct tribe in *The Storyteller*, and Peter Matthiessen's missionary-adventure story in which a North American man falls in love with an Indian woman, only to die from the fatal disease transmitted by her kisses: a classic incorporation of ambivalence.[106] The television producer Alan Ereira describes how he discovered a tribe of Indians lost for four centuries who had retreated up a mountain from the Spanish to create an ecologically sound communal way of life, which they now believe, because the ice-cap has melted from their mountain, it is their duty to give to the outside world, whom they term their irresponsible 'younger brother'.[107] The same tone infects contemporary media debate about the environmental threats to the Amazon and its peoples.[108] Anthropologists again sing similar tunes. Darcy Riberio's Amazonian Indian love story *Maíra* has proved a big popular success; and a recent book by Marcos Santilli, with its text from both Indians and anthropologists in English as well as Portuguese, includes exquisitely subtle portraits of nude young Indian women in a portrait of the forest and its people. Its title, *Àre*, means 'brothers, companions' in the Suruí language, and the book is essentially a celebration of one of the world's last societies still living 'isolated in forests, in small self-sufficient societies, without crime, without madness, without sexual conflicts, without inequalities, without exploitation, without oppression; [living in] coexistence with wood, adaptation to nature, co-operation, spiritual evolution'.[109] The last wild people, innocent, sexually expressive. Such dreams may be in the minds of many of those who set out, however seriously, for Brazil.

Some north European men come as 'sex tourists', seeking the sun, the beaches, and 'summer love' with brown-skinned women whom they believe to be vitally erotic.[110] Others are more attracted to contact with the forest people, to find spontaneous innocents still uncorrupted by western capitalist life. They may imagine even hearing tales of the strange reptilian god Curupira, with green teeth and green feet, or the secret oral traditions of *The Storyteller*.[111] Yet others are missionaries of the new environmentalist movements, seeking to protect the Amazon's infinite reserves of undiscovered species, animals, birds, insects, flowers, vital organisms, the world's breathing space. And why do these seekers of Eden converge here, rather than in the rainforests of Africa or Asia? It seems as if the duality of the wild men has finally split. Wild men have disappeared from imaginations of Asia, the continent of old civilizations. And all that is theatening is now concentrated in Africa, the 'dark continent' of the Victorians, Conrad's *Heart of Darkness*, today's continent of famine, civil war, tribal massacre and genocide. Europeans look to the Americas for hope: for mankind and for the environment.

The rise of environmentalism has been the last great force of change transforming the image of the wild and the green: and it is this above all which has offered a powerful new image which fuses the wild and the green man. William Morris is now seen as the first great Engish prophet of the green movement: a socialist who understood how the driving force of capitalism was destroying both the diversity of human cultures and the natural environment of mankind.[112] A hundred years later, for north Europeans, faced with runaway urbanization and motorization, threatened by drastic climatic changes, seeking spirituality in a material secular era, environmentalism has become a crusade which unites people of all ages and backgrounds: from the conservative elderly well-to-do to the workless radical young.

This fusion of the wild and the green is both diffuse and eclectic. You stumble upon it in many unexpected places as well as where consciously intended. You can hear it when your train is late in the autumn because of 'leaves on the line', and the French manager of a privatized British Rail service blames delays on our 'culture of viewing trees as "sacred"' and so not cutting sufficient trees close to the track.[113] You can find it in strange health innovations, like the Dryad-watched *Tree Wisdom* of Jacqueline Memory Paterson; or the psychotherapeutic *Women Who Run with the Wolves: Contacting the Power of the Wild Woman* – the feminine soul – by Clarissa Pinkola Estés, Mexican-Spanish feral child of Wyoming;[114] in the news that the French have developed a new cure for impotency based on organic extracts from 'a Brazilian shrub known locally as Potency Wood';[115] or the display in a London 'Ecoll' shop window of a 'natural' alternative to HRT, prolonging women's sexual vitality, long known to Brazilian and Mexican women, and now packaged by Vitality of East Grinstead. And in another way in the beautiful evanescent Scottish tree and leaf sculpture of Andy Goldsworthy.[116] Or in the inauguration at Clun on Bank Holiday Monday 1997 of an entirely new Green Man Day, sponsored by the local Tourist Board and organized by the retired managing director of Legoland, who had himself created as its innovative symbol a green man foliate head with deer antlers. Perhaps to add a touch of respectability, the day started with a historical talk by a bearded lecturer who argued that through our DNA we all carry ineradicable genetic memories of the Druidic green man, our timeless fertility symbol. Meanwhile outside in the street the crowds gathered to greet the procession down from the church: the green man followed by Mayday children and Morris dancers. At the town bridge Winter challenged the Green Man, but was quickly blown aside. And then the fair, with stalls for plants, green man sculptures and T-shirts and paintings, home-made jam and Women's Institute cakes. In small-town rural Clun, to espouse the myth of greenness looked very like accepting a repackaging of what was there already.

With such continuing adaptive flexibility, it would seem unlikely that the myths of the wild and the green will not outlast us. But what of the environment itself? A mere repacking will no longer suffice to protect it: for survival, humanity needs a radical rethinking of priorities. And fortunately that rethinking has its missionaries.

So for the future the focus must be above all on the new young city militants. It is these last who are the new wild people in England, the movement's martyrs, homeless campaigners against destructive road-builders who pit their bodies against bailiffs and bulldozers. These visionary green men and women, some like troglodytes taking refuge in underground tunnels, others like clusters of desert anchorites living in the treetops, standard-bearers of old traditions, are the new prophets of the wild and the green.

Notes

I am particularly grateful for encouragement and help given to me by Bill Anderson, who sadly did not live to see this text; for moral support given to me by Tony Woodiwiss, my Head of Department, who himself claims descent from a woodwose; for advice given to me by Jonathan Alexander, Natasha Burchardt, Giovanni Contini, Iris Gareis, Diana Gittins, Catherine Hall, Roger Peers, Adriana Piscitelli, Cesare Poppi, Sandro Portelli, Rumi Sakamoto, Jacqueline Sarsby, Revan Schendler, Richard Candida Smith, Norman Talbot, John Tchalenko; and especially to Andréa Zhouri, whose current research, on British campaigners working to protect the Amazon tropical forests, first drew my own interest to these questions.

1 K. Thomas, *Man and the Natural World* (London, 1983), 193.
2 I am grateful to Revan Schendler for giving me this proverb.
3 Information from the Woodland Trust, 1 August 1997; A. Zhouri, 'Pathways to the Amazon: British Environmental Campaigners', *Communicating Experience*, IX International Oral History Conference, June 1996, Göteborg, vol. 2, 284–91. An opinion poll of 1,000 adults, carried out for the Woodland Trust in 1998, reported that two thirds of British adults were concerned about the fate of threatened woods in Britain, and 90 per cent consider ancient woodland to be as important to national heritage as castles and cathedrals.
4 S. Hecht and A. Cockburn, *The Fate of the Forest* (London and New York, 1989); W. Anderson, *Green Man* (London, 1990), 160.
5 G. Harvey, *The Killing of the Countryside* (London, 1997), 7; 97 per cent of meadowlands were also lost in the post-war period (12).
6 Thomas, *Man and the Natural World*, 193–4.
7 A. Trollope, *The Way We Live Now* (London, 1874), 328.
8 Hecht and Cockburn, *The Fate of the Forest*, 6–8.
9 L. Spence, *The History and Origins of Druidism* (London, 1949), 11, 78.
10 Thomas, *Man and the Natural World*, 214–15; Anderson, *Green Man*, 50–1.
11 T. Pakenham, *Meetings with Some Remarkable Trees* (London, 1996), 96–101.
12 This was first observed by Giovanni Contini, now a regional social historian, who grew up below the mountain, but only realized the significance of the earthworks in the wood after visiting the Gog Magog hillfort outside Cambridge.
13 *New Shorter Oxford English Dictionary*; G. de Champeaux and Dom S. Sterkx o.s.b., *Le Monde des symboles* (Paris, 1989), 306.
14 R. Lannoy, *The Speaking Tree* (London, 1971), xxv.
15 T. Weber, *Hugging the Trees: The Story of the Chipko Movement* (New Delhi, 1988).
16 De Champeaux and Sterkx, *Monde des symboles*, 365–73.

17 This myth was especially popular in sixteenth- to seventeenth-century Italy: famous depictions include Antonio del Pollaiuolo's painting in the National Gallery, London, and Bernini's sculpture in the Villa Borghese, Rome. Later it was taken up by William Blake, and by many early-twentieth-century artists. I am grateful to Elaine Bauer for drawing my attention to this theme.

18 An outstanding architectural example is the Horniman Museum in Forest Hill, South London, by Harrison Townsend (1901), with a tower festooned in tree carvings.

19 I. Nicholson, *Firefly in the Night: A Study of Ancient Mexican Pottery and Symbolism* (London, 1959).

20 Genesis 2.9.

21 A particularly fine example is the Adam from Notre Dame, Paris, now in the Musée Cluny.

22 De Champeaux and Sterkx, *Monde des symboles*, 321–3.

23 Thus William Bromflet, carver of the Ripon misericords in the 1490s, which include a wild man, was later city mayor; while Flemish sculptors were lured to work in both France and Scotland: N. Coldstream, *Medieval Craftsmen: Masons and Sculptors* (London, 1991); C. Grössinger, *Ripon Cathedral Misericords* (Ripon, 1989); D. and H. Kraus, *The Hidden World of Misericords* (London, 1976); L. Maeterlinck, *La Genre satirique, fantastique et licencieux dans la sculpture flamande et wallonne: les miséricorde de stalles (art et folklore)* (Paris, 1910).

The green man image also occurs on the thirteenth-century painted nave ceiling of Peterborough cathedral; there may have been other instances of painters using the image, now lost.

24 D. Bruna, *Enseignes de pèlerinage et enseignes profanes* (catalogue, Musée Cluny, Paris, 1996): the emblems were dredged up from the mud of the Seine and the Scheldt.

25 Winchester was a medieval royal capital and its thirteenth-century great castle hall has the magnificent contemporary round table of the knights, now hanging on its end wall.

Despite his first comment, Anderson later convincingly describes the twelfth-century naked men and women in foliage at Brioude, Auvergne, as mermen and mermaids rather than green men and women. These are half-length figures. The fifteenth-century Ulm green woman is not illustrated in full: *Green Man*, 23, 66–7, 101–2. There are a number of English medieval foliate heads which look to be possibly female, for example in Tewkesbury Abbey and Norwich Cathedral, but I know of no definite instances. C. J. P. Cave found some green woman bosses, but describes them as 'much rarer': *Roof Bosses in Medieval Churches* (Cambridge, 1948), 65.

26 Anderson argues that during the Dark Ages combinations of face masks and interlaced vegetation are found especially in Ireland, then cut off from the Roman church by the Saxon invasions – as in the Book of Kells or the cross at Clonmacnois: *Green Man*, 45–6, 55. However his illustrated examples are not convincing, and it is certain that despite Celtic fascination for interlacing foliage the foliate head was rarely if ever used as one of its stock of visual images: L. and J. Laing, *Art of the Celts* (London, 1992).

27 K. Basford, *Green Man* (Ipswich, 1978), 7.

28 Anderson, *Green Man*, 105.

29 As in roof bosses at Blyth and New Shoreham, where the foliage makes the men's heads invisible from below: Cave, *Roof Bosses*, 65–6.

30 The master had gone to Rome for advice, but the apprentice learnt how to carve the pillar from a dream: *Guardian*, 18 February 1997; Anderson, *Green Man*, 124–7. On the masonic tradition, see R. Brydon, *Rosslyn: A History of the Guilds, the Masons and the Rosy Cross* (Roslin, 1994).

31 N. Pevsner, *The Leaves of Southwell* (Harmondsworth, 1945); N. Summers, *The Chapter House, Southwell Minster* (Derby, 1994).
32 C. J. P. Cave, *Medieval Carvings in Exeter Cathedral* (Harmondsworth, 1953).
33 Thomas, *Man and the Natural World*, 194.
34 R. Bernheimer, *Wild Men in the Middle Ages* (Cambridge, MA, 1952), 98.
35 Especially R. Bartra, *Wild Men in the Looking Glass* (Ann Arbor, MI, 1994); Bernheimer, *Wild Men in the Middle Ages*; R. Togni, 'L'Uomo selvatico nelle imagine artistiche e letterarie', *Annali di San Michele* (Trento), 1 (1988), 88–154. By contrast there is only one major study of the green man (Anderson, *Green Man*) and while immensely knowledgeable this is somewhat mystical in its perspectives.
36 Genesis 25.7.
37 Bartra, *Wild Men in the Looking Glass*, 45.
38 A striking example is by Donatello at the Bargello, Florence.
39 J. Tchalenko, 'Earliest Wild-man Sculptures in France', *Journal of Medieval History*, 16 (1990), 217–34.
40 R. Wittkower, 'Marvels of the East: A Study in the History of Monsters', *Journal of the Warburg and Courtauld Institutes*, 5 (1942), 159–97; J. Block Friedman, *The Monstrous Races in Medieval Art and Thought* (Cambridge, MA, 1981).
41 This is likely also to be the meaning of the small naked but hairless man sitting underneath the great pulpit by Nicola Pisano in the baptistery of Pisa cathedral (1260), who has sometimes been called a wild man.
42 Bernheimer, *Wild Men in the Middle Ages*, 2, 71; Bartra, *Wild Men in the Looking Glass*, 179.
43 Bernheimer, *Wild Men in the Middle Ages*, 36.
44 Togni, 'L'Uomo selvatico', 90; Bartra, *Wild Men in the Looking Glass*, 45; O. Mazur, *The Wild Man in the Spanish Renaissance and Golden Age Theater: A Comparative Study Including the Indio, the Barbaro and their Counterparts in European Lores* (University Microfilms International, Ann Arbor, MI, 1980).
45 Bernheimer, *Wild Men in the Middle Ages*, 33–8.
46 J. Alexander, 'Dancing in the streets', *Journal of the Walters Art Gallery*, 54 (1996), 147–62.
47 Bernheimer, *Wild Men in the Middle Ages*, 30.
48 Bartra, *Wild Men in the Looking Glass*, 105; Togni, 'L'Uomo selvatico', 108 (in Albertina, Vienna).
49 Bernheimer, *Wild Men in the Middle Ages*, 39, 105–6; Museum für Kunst und Gewerbe, *Die wilden Leute des Mittelalters* (exhibition catalogue, Hamburg, 1963), 33; Togni, 'L'Uomo selvatico', 106–7 (tapestry from Strasburg, now in Museum of Fine Arts, Boston); T. Husband, *The Wild Man: Medieval Myth and Symbolism* (New York, 1980).
50 Bartra, *Wild Men in the Looking Glass*, 90; Bernheimer, *Wild Men in the Middle Ages*, 33.
51 Bernheimer, *Wild Men in the Middle Ages*, 34, 121–75; *The German Legends of the Brothers Grimm*, translated by D. Ward (London, 1981), vol. I, 52–3; Bartra, *Wild Men in the Looking Glass*, 97. Mary Magdalene's sexuality was also typically conveyed through her very long hair. Her hair almost touches her ankles in the della Robbia majolica of her return from the desert, now in the Museo del Opera del Duomo, Florence.
52 To Bishops Godfrey and Frederick von Bülow, 1375, Schwerin, Germany: M. Norris, *Brass Rubbing* (London, 1965), 72–3, 80–1.
53 Bernheimer, *Wild Men in the Middle Ages*, 40; Grimm, *German Legends*, vol. I, 62.
54 Bartra, *Wild Men in the Looking Glass*, 100.
55 For example, the mermaid roof boss at Sherborne Abbey, or the misericord at Lincoln Cathedral.

56 Grimm, *German Legends*, vol. I, 54; H. Andersen, *Folk Tales* (London, 1988), 205–29 (beautifully illustrated by Arthur Rackham in 1932).

57 F. J. Stopp, 'Henry the Younger of Brunswick: Wild Man and Werwolf in Religious Polemics, 1538–1544', *Journal of the Warburg and Courtauld Institutes*, 33 (1970), 200–34; Anderson, *Green Man*, 134–5 (Luther's Appeal to the General Council at Wittenburg, 1520; painting of Luther preaching by Cranach the Elder); Bartra, *Wild Men in the Looking Glass* (engraving by Melchior Lorsch to text by Luther, 1545).

58 At Lübeck the wild man is an intentional counterpart to a figure of the Emperor; at Venice the iconography is not typical, but he is perhaps intended as a feral adolescent. (This carving was hidden when Ruskin carried out his analysis, so is not included by him as part of the life cycle carvings on the other sides of this north-eastern capital of the lower arcade, facing the sea.)

59 C. Wood, *Albrecht Altdorfer and the Origins of Landscape* (London, 1993), 86–7, 152–8.

60 Anderson, *Green Man*, 136–9.

61 Similar continuities can also be found in Flanders, including Lille, where artisan traditions were again very strong.

62 The few possible instances I know in Burges' work are curiously classical in feeling, such as the masks on the screen in Cork cathedral, or Lord Bute's claret jug: J. Mordaunt Crook, *William Burges and the High Victorian Dream* (London, 1981), plate 258. Surprisingly, one example of an unmistakably gothic foliate head is to be found on the St Marylebone Grammar School, London, by the unscholarly architect E. Habershon, 1856. Another, more tentative, apparently unique and not imitated, is a headstop on the south-west porch of St Augustine's, Ramsgate, by Pugin himself, 1850.

63 Anderson, *Green Man*, 151–3, gives examples in Mayfair, London, of the 1890s, and I have found other similar instances elsewhere from the same period, but almost all are ambiguous, and much more likely to be intended as satyrs. Surprisingly, one of the few buildings of the period which is profusely sculptured with undeniable green men is the Ontario Parliament Building in Toronto, Canada, by Richard Waite (1892). Waite was an English-born architect who had been working in Buffalo, and had imbibed strong influences from both H. H. Richardson and Louis Sullivan. Another is on the plaster coving of the house of the woodcarver J. E. Elwell at 45 North Bar Without, Beverley, Yorkshire (1894).

64 Lady Raglan, 'The "Green Man" in Church Architecture', *Folklore*, 50 (1939), 45.

65 Raglan, 'The "Green Man"', 50.

66 R. Hutton, *Stations of the Sun* (Oxford, 1996), 226.

67 Bernheimer, *Wild Men in the Middle Ages*, 61–71; Anderson, *Green Man*, 12, 133–4; B. S. Centerwall, 'The Name of the Green Man', *Folklore*, 108 (1997), 25–33; Matthew Taubman, *London's Yearly Jubilee* (London, 1686).

68 J. Strutt, *The Sports and Pastimes of the People* (London, 1801), vol. IV, 261, 282: he copied his illustrations of the green man from the title page of a much earlier firework manual; here, however, the figure was not described as a green man: J. Bate, *The Mysteries of Nature and Art* (London, 1635).

69 D. Kennedy, *England's Dances* (London, 1949), 49–50.

70 R. Judge, *The Jack in the Green* (Ipswich, 1979); Hutton, *Stations of the Sun*.

71 Centerwall, 'The Name of the Green Man', 27, 32; B. Lilywhite, *London Signs* (London, 1972). Herbalists were known in the late middle ages as 'green men' because they dealt in green herbs, but Green Man pubs only appeared in the seventeenth century.

72 Thomas, *Man and the Natural World*, 15, 194, 198. Early medieval instances of treeplanting are given by J. Harvey, *Medieval Gardens* (London, 1981), 13–17.

73 Thomas, *Man and the Natural World*, 178–9, 193, 213–14; 'Binsey Poplars', *Gerard Manley Hopkins* (Harmondsworth, 1953), 39.

74 Thomas, *Man and the Natural World*, 216; R. W. Emerson, *Works* (London, 1902), 548; J. Muir, *Nature Writings* (New York, 1997), 629.

75 W. Morris, *Letters*, vol. I, 377–8; vol. II, 515; *Collected Works*, vol. XV, *The Roots of the Mountains*, 20–1, 75; vol. XX, *The Water of the Wondrous Isles*, 15–17, 335; N. Talbot, 'William Morris and the Wildwood' (unpublished manuscript, 1996).

76 E. Thompson Seton, *Two Little Savages: Being the Adventures of Two Boys Who Lived as Indians, and What They Learned* (London and New York, 1903); E. Thompson Seton, *The Birchbark Roll of the Woodcraft Indians* (London and New York, 1906), retitled in later editions *The Woodcraft Manual*. The idea (discarded by the breakaway socialist Woodcraft Folk founded in 1925) of 'chivalry' suggests how Seton was drawing his ideas from a Gothic Revival tradition: earlier antecedents include not only Morris, but also Sir Walter Scott's notion of 'woodcraft', for example in *The Bride of Lammermuir* (Edinburgh, 1830), chapter 3.

77 E. Westlake (ed.), *The Order of Woodcraft Chivalry* (London, 1918); E. Westlake, *The Forest School: The Principles of Education of the Order of Woodcraft Chivalry* (Fordingbridge, 1927), British Library copy destroyed by wartime bombing; Anon., *Our Story: Fifty Years under Canvas with Forest School Camps* (Crickhowell, 1998).

78 *The Famous History of Valentine and Orson, or the Wild Man of Orleans, Containing an Account of All the Wonderful Adventures, and Surprising Achievements, of these Two Valiant Brothers, with the Affecting Distresses of their Unfortunate Mother, the Fair Bellisant, Empress of Constantinople* (London, 1804); *Valentine and Orson, a Tale* (Birmingham, 1810).

79 *Authentic Anecdotes of the Wild Man of the Woods, Found by King George the Second on a Hunting Party in Germany, and Many Years Known Here by the Name of Peter the Wild Boy* (London, 1820); *Peter the Wild Boy: An Enquiry how the Wild Youth lately Taken in Woods near Hanover (and Now Brought to England) Could Be Left There and By What Creature He Could Be Suckled, Nursed, and Brought Up* (London, 1726); *The Devil to Pay at St James: A Full and True Account of . . . How the Wild Boy is Come to Life again, and Has Got a Dairy Maid with Child . . .* (London, 1727); *It Cannot Rain but It Pours: Or, London Strow'd with Rarities; Being an Account . . . of the Wonderful Wild Man that was Nursed in the Woods of Germany by a Wild Beast . . . and is a Christian like One of Us, Being Call'd Peter; and how He Was Brought to Court all in Green . . .* (London, 1726); *La Belle Sauvage: The True and Surprising History of a Strange Girl Found Wild in the Woods of Champaigne* (London, 1820). A contemporary image of Peter the Wild Boy among the royal courtiers can be seen in the staircase murals of Kensington Palace (by William Kent, 1725). However, he is fully clothed and, like many young boys of that time, wears a dress; I noticed that attendants incorrectly identify him to present-day visitors as the figure of a black servant boy.

80 J.-M.-G. Itard, *De l'Éducation d'un homme sauvage, ou, Des Premiers développements physiques et moraux du jeune sauvage d'Aveyron* (Paris, 1801); J.-M.-G. Itard, translated by G. and M. Humphrey as *The Wild Boy of Aveyron* (London, 1972).

81 J.-J. Rousseau, *Discours sur les sciences et arts* (Dijon, 1750); E. Dudley and M. Novak (eds), *The Wild Men Within: An Image in Western Thought from the Renaissance to Romanticism* (London, 1972); H. N. Fairchild, *The Noble Savage: A Study in Romantic Naturalism* (New York, 1928); R. Gonnard, *La Légende du bon sauvage: contribution à l'étude des origines du socialisme* (Paris, 1946); Mazur, *Wild Man in the Spanish Renaissance*.

82 Bernheimer, *Wild Men in the Middle Ages*, 87, 93; Togni, 'L'Uomo selvatico', 143. Another English instance is a woodcut depicting an aborigine as a wild man in J. Bulwer, *Anthropometamorphosis: Man Transform'd: or, the Artificial Changeling* (London, 1650).
83 Thomas, *Man and the Natural World*, 194.
84 Bartra, *Wild Men in the Looking Glass*, 1.
85 Bartra, *Wild Men in the Looking Glass*, 171–3. Shakespeare owned a copy of Montaigne's *Essais*: H. Honour, *The New Golden Land* (London, 1975), 62.
86 Especially in *El Premio de la Hermosura*; for a detailed discussion, see Mazur, *Wild Man in the Spanish Renaissance*.
87 It remains a mystery why they decorate the tomb of Edward Harman, but he had connections with the royal court, and one must assume that his enterprises involved trade with South America.
88 Honour, *New Golden Land*, 6–11. The legendary arrival scene is recycled in Oscar Curtino's 1996 painting, 'Cristobal Colon', Latin American Art Collection, University of Essex.
89 For example, anthropologists such as Margaret Mead wrote confidently about cannibalism without ever witnessing it, and it is possible that even 'savage' ritual cannibalism went no further than the symbolic sacrifice practised regularly by Christians: W. Arens, *The Man-Eating Myth* (New York, 1979); W. Arens, 'Man is Off the Menu', *Times Higher Education Supplement*, 12 December 1997.
90 R. Hamor, *A True Discourse of the Present State of Virginia* (London, 1615), 1.
91 For example in the paintings of George Catlin in the National Gallery of American Art, Washington, and of Paul Kane in Toronto; see also W. H. Truettner, *The West as America: Reinterpreting Images of the Frontier* (Washington, 1991), 151ff.
92 R. M. Ballantyne, *The Wild Man of the West: A Tale of the Rocky Mountains* (London, 1863) (Ballantyne was most famous as author of *Coral Island*; visually there are also distinctly Arthurian hints in the first edition); J. Blaine, *Wild Man at Smoke Creek* (London, 1972). One best-selling Australian parallel, of uncertain authorship, was *Wild Colonial Boys* (Melbourne, 1948): see Munro's biography of *Inky Stephenson: Wild Man of Letters* (Melbourne, 1984), who was the ghost writer of the official author, Frank Clune.
93 I owe this point to Sandro Portelli.
94 *Kentucky's Family Tree*, Kentucky History Center, 1996.
95 Exhibition catalogue, *Frederick E. Church: Under Changing Skies*, Arthur Ross Gallery (University of Pennsylvania, 1992); there is a permanent display of Church's paintings in his own house, Olana, Hudson, New York.
96 H. D. Thoreau, *Walden* (Boston, 1854: Harmondsworth, 1983), 182–3, 303–9.
97 N. Weekly, *Charles E. Burchfield: The Sacred Woods* (New York, 1993), 15–16, 70. This vision of the forest as both a challenge for survival and mystical can be found continuing in a successful modern Penguin paperback American story for younger children by T. Locker, *Where the River Begins* (New York, 1993).
98 J. Dryden, *The Conquest of Granada* (London, 1672).
99 W. Wordsworth, *Prelude*, I, 11, 297ff; Honour, *New Golden Land;* Fairchild, *Noble Savage*.
100 Fairchild, *Noble Savage*, 104–11; exhibition, National Maritime Museum, London, 1997.
101 E. B. Berthet, *L'Homme des bois*, translated as *The Wild Man of the Woods: A Story of the Isle of Sumatra* (London, 1868); E. Mershon, *With the Wild Men in Borneo* (Mountain View, CA, 1979); J. Greenwood, *The Wild Man at Home: Or, Pictures of Life in Savage Lands* (London, 1879) (e.g. the Borneo snake dance, 141).
102 See his woodcuts, from his *fauve* ('wild beast') phase, illustrating Guillaume Apollinaire's first book, *L'Enchanteur pourristant* (Paris, 1909).

103 M. Mead, *Coming of Age in Samoa* (New York, 1928).
104 J. Reid, *Oxford Guide to Classical Mythology in the Arts, 1300–1990* (Oxford, 1993); for Cosin, Christie's sale catalogue, *Old Master Drawings*, 16 April 1997.
105 W. H. Hudson, *Green Mansions: A Romance of the Tropical Forest* (London, 1904; a subsequent 1926 edition was beautifully illustrated by Keith Henderson); Hecht and Cockburn, *Fate of the Forest*, 11–13; A. Russell Wallace, *Narrative of Travels on the Amazon and River Negro, with an Account of the Native Tribes* (London, 1853); J. B. Debret, *Viagem Pitoresca e Histórica ao Brasil* (São Paulo, 1954); J. M. Rugendas, *Viagem Pitoresca Através do Brasil* (São Paulo, 1967); C. Slater, 'Amazonia as Edenic Narrative', in W. Cronon (ed.), *Uncommon Ground: Toward Reinventing Nature* (New York and London, 1995), 219; T. Roosevelt, *Through the Brazilian Wilderness* (New York, 1914).
106 M. Rome, *The Wild Man* (London, 1980); M. Vargas Llosa, *The Storyteller* (London, 1990); P. Matthiessen, *At Play in the Fields of the Lord* (New York, 1965); also A. Carpentier, *The Lost Steps* (Harmondsworth, 1968).
107 A. Ereira, *The Heart of the World* (London, 1990).
108 Zhouri, 'Pathways to the Amazon'; Slater, 'Amazonia as Edenic Narrative'.
109 D. Riberio, *Maíra* (New York, 1984); M. Santilli, *Àre* (São Paulo, 1987), 15–16.
110 A. Piscitelli, '"Sexo Tropical": Comentários sobre gênero e "raça" em alguns textos da mídia brasiliera' (unpublished paper, 1996).
111 M. Vargas Llosa, *The Storyteller*.
112 P. Thompson, *Why William Morris Matters Today: Human Creativity and the Future World Environment* (London, 1991).
113 A. Hurel, manager of Connex, *Evening Standard*, 21 November 1997.
114 J. Memory Paterson, *Tree Wisdom: the Definitive Guidebook to the Myth, Folklore and Healing Power of Trees* (London and San Francisco, 1996); C. Pinkola Estés, *Women Who Run with the Wolves: Contacting the Power of the Wild Woman* (London, 1992).
115 *Evening Standard*, 16 September 1998.
116 A. Goldsworthy, *Wood* (London, 1996).

Select bibliography

Anderson, W., *Green Man: The Archetype of Otherness with the Earth* (London, 1990).

Bartra, R., *Wild Men in the Looking Glass: The Mythic Origins of European Otherness* (Ann Arbor, MI, 1994).

Basford, K., *Green Man* (Ipswich, 1978).

Bernheimer, R., *Wild Men in the Middle Ages: A Study in Art, Sentiment and Demonology* (Cambridge, MA, 1952).

Cave, C. J. P., *Roof Bosses in Medieval Churches* (Cambridge, 1948).

Coldstream, N., *Medieval Craftsmen: Masons and Sculptors* (London, 1991).

Dudley, E. and Novak, M. (eds), *The Wild Man Within: An Image in Western Thought from the Renaissance to Romanticism* (London, 1972).

Fairchild, H. N., *The Noble Savage: A Study in Romantic Naturalism* (New York, 1928).

Frazer, Sir J. G., *The Golden Bough: A Study in Magic and Religion* (abridged edition, London, 1992).

Friedman, J. B., *The Monstrous Races in Medieval Art and Thought* (Cambridge, MA, 1981).

Harvey, G., *The Killing of the Countryside* (London, 1997).

Hecht, S. and Cockburn, A., *The Fate of the Forest: Developers, Destroyers and Defenders of the Amazon* (London and New York, 1989).

Honour, H., *The New Golden Land: European Images of America from the Discoveries to the Present Times* (London, 1975).

Husband, T., *The Wild Man: Medieval Myth and Symbolism* (New York, 1980).

Judge, R., *The Jack in the Green: A May Day Custom* (Ipswich, 1979).

Lannoy, R., *The Speaking Tree: A Study of Indian Culture and Society* (London, 1971).

Raglan, Lady, 'The "Green Man" in Church Architecture', *Folklore*, 50 (1939), 1, 45–57.

Thomas, K., *Man and the Natural World: Changing Attitudes in England, 1500–1800* (London, 1983).

Togni, R., 'L'Uomo selvatico nelle immagini artistiche e letterarie: Europa e arco alpino (secoli XII–XX)', *Annali di San Michele* (Trento), 1 (1988), 88–154.

Weber, T., *Hugging the Trees: The Story of the Chipko Movement* (New Delhi, 1988).

Wittkower, R., 'Marvels of the East: A Study in the History of Monsters', *Journal of the Warburg and Courtauld Institutes*, 5 (1942), 159–97.

2 Animals, children and peasants in Tuscany

A note on the San Gersolé archive

Giovanni Contini

A hierarchical and hard world

In Italy as elsewhere peasant society, following its decline and disappearance, has been retrospectively romanticized not only popularly, but also by many historians and anthropologists.[1] One particularly widespread view is that peasants lived in a symbiotic mutually harmonious relationship with the natural world, not only depending on the fruitfulness of nature but caring for it and conserving it. But the reality could be altogether otherwise.

A vivid illustration of this is to be found in the collection of notebooks written from the 1930s to the 1950s by peasant children in the small village of San Gersolé, near Florence.[2] This is an extremely important archive for studying the culture of the Tuscan peasants in the last years of the sharecropper system, the *mezzadria*, which came to an end by the late 1950s. Because their exceptionally far-sighted teacher Maria Maltoni had based her teaching on the observation and description of the surrounding reality, the children's writings provide a rare document of everyday peasant family life from the inside: the result is a fascinating collection of materials which are, in a sense, like oral documents transcribed by their own authors. The young writers used their everyday language in order to describe the natural world, and they also often include verbatim everyday dialogues between the adults in their circle.

There is little romanticization of the peasant world here. On the contrary, the children instead portray it as a very violent and hierarchized world: a social pyramid which had the landlord at the top, followed, moving downwards, by the bailiff, the peasant family's leader (the *capo di casa* or *capoccia*), then the men of the family, the women, the male children, the female children. Below them we find the daily labourers (*pigionali*, from *pigione*, rent: those who were forced, unlike the sharecropper peasants, to rent a house). Then the variegated world of the outcast: pedlars, ragmen, mendicant friars and nuns, beggars, gypsies. Last, even lower still, came the natural world of domestic and savage animals.

The relationship between the inhabitants of this world depended on the social position of each member. The landlord, a remote reality, held the reins of power. On the very rare occasions when he (or she) approached a peasant family directly, the landlord expected to be addressed with a humble deference. He was entitled to reproach every member of the family about a whole variety of everyday matters: 'the threshing-floor around the house is not clean enough', 'too many people are idle at home in the middle of the workday', 'the field must be weeded', and so on. He was also entitled to enter the house to inspect, and to give notice to quit, if the family had made some important mistake, such as stealing some of the landlord's crop share, or even marrying a member of the family without permission from the bailiff, or showing too expensive clothes at the Sunday mass.

As in every traditional society based on deferential/paternalistic relationships, the landlord complemented this fundamentally repressive role with a more 'human', 'paternal' one: he could give presents to the children, second-hand clothes to the women. He could give special help to reward a particularly 'deserving' family, loyal and religious, or to a family suffering from the economic burden of too many children to feed, 'mouths' as yet incapable of work.

Even the peasant family itself had a very hierarchical structure. The leader, always a man (often unmarried in order to avoid favouritism), was informally 'elected' by the family council, which consisted only of men. Once elected, he was the exclusive representative of the family to the landlord; he had exclusive charge of the family's cash; he was the only one who left the house and the field to take part in the local markets. There are many anecdotes about the sumptuous dinners around the market square of these peasant family leaders, not to speak of their dalliances with prostitutes . . .

This exclusively male family council also took decisions on all the most important family questions. The women of the family were thus excluded from important decisions such as whether to leave the plot for a better one or not, how to invest the savings of the family, whether to buy or sell cows, horses, donkeys, and so on.

Even among the women there was a leader (the *massaia*: sometimes the wife of the *capo di casa*). The *massaia* had the keys of the storeroom, and the *capo di casa* gave her money in order to buy food, clothing and other necessary articles for all the members of the family. Women in the *mezzadria* region also had to perform quite heavy jobs in the fields, in addition to the traditional housework. Yet the only money which they had at their own disposal came from the little which they could manage to raise from the *bassa corte*, through selling eggs, chickens or rabbits.

Finally, at the bottom of this hierarchical pyramid, we find the children themselves, the writers of these school notebooks.

Adults seem to have given as little time as possible to bringing their children up, and in explaining what was right and wrong behaviour; why it was possible to do something, and impossible to do something else. Children had simply to follow the adults' example. If they failed to do so, they quickly learnt that a verbal reproach was likely to be immediately followed by blows.

Children did not always accept this kind of discipline easily. Thus in November 1937 Fernanda Caroti unintentionally killed a rabbit. Her father beat her. Hurt and angered, she said to him, 'I hope you'll never recover from your back illness'; and was rewarded by still more blows. A boy of the same time, Natalino Carrai, often wrote of successfully running away to escape some adult's assault. In February 1938 he described pretending to push a heavy cart, and so managing to escape his father whom he told, 'I'm not afraid', his father retorting, 'I'll manage to scare you.'

Ten years later Marcella Pampaloni had an altercation with her father, at the end of which she calls him *bischero* (a word which means tuning peg, yet has a second meaning, widespread in Tuscany, as a sexual metaphor). Both her parents caught her and her father whipped her so hard that the marks were still raw a week later. On a later occasion when she was working in the field, Marcella suddenly threw away her sickle saying, 'This job doesn't suit me.' Her uncle turned to her father and said, 'Look how she behaves' ('guarda che verso e la fa'): an idiomatic expression which contains an allusion to the misbehaviour of a domestic animal. Meanwhile her father was preparing to whip her again.

Children and domestic animals

When the children describe their relationships with domestic animals, these tend to repeat those between their parents and themselves, for the animals were often treated very roughly when they misbehaved.

Cats were still considered as half-wild animals. When Marcella Pampaloni's father once found the cat sleeping in her bedroom, he reacted so badly that the girl tried to justify herself by saying that the animal had been misled by an old lady from Florence who had found shelter from the bombs in the city in the Pampaloni's house (the diary was written during the war). In general, the cat was not fed, but was supposed to find its own food (mice and other vermin) by itself. If it stole meat from the kitchen, or if it killed a chick or chicken, the cat's punishment could be death.

But it could also be killed just for fun. Natalino Carrai waged a permanent war against cats: he stoned them, he drowned their kittens in the river. Once Natalino and a cousin tied the cat to the family's cart, then 'I hammered it in the back with a club . . . then we saw it laying down in the path, half dead' ('gli ho tirato una palucciata nel groppone, . . . si è visto a giacere nel mezzo di un viottolino mezzo morto'). His sister merely made the mild comment that the cat had been good at catching mice.

Even other more important domestic animals, such as hens or rabbits, could be killed simply for fun. Once Fernanda Caroti hurt a hen with a stick, and before its death, in order to protect herself from punishment by the adults, she crushed its comb so that it looked as if a cock had attacked it ('gli stiacciai la cresta come che l'avessero beccata i polli'). The teacher was shocked, and wrote in Fernanda's notebook: 'How nasty, Fernanda, that you dared to crush the comb when she was still alive!'

But the girl did not feel guilty. She, like other children too, often stoned the chickens, sometimes killing them. They would also beat cats to teach them not to eat young chicks, or tied them up in sacks and carried them off, leaving the unfortunate cats far away from home, or worse still, throwing them into a river.

Children and the wilderness

The edges of the peasant holding or the woods were the places where children would seek shelter from the control and punishments of adults. But the wood was also a playspace during more normal times, particularly for the boys. Here they would spend hours hunting the wild birds and small animals. Such hunting was indeed the favourite hobby of very many children. In a group of boys Natalino Carrai and his friends set numerous traps in the nearby countryside in order to catch small birds. Paradoxically however, it seems that for these children a favourite sport was not so much catching the birds with their own small traps, but rather stealing each other's traps.

Natalino and the others not only caught the adult birds, but they also plundered their nests. They seem to have been extraordinarily expert in detecting where the nests were and in managing to reach them even if they were very high up in the trees. Through their egg-stealing they were able to identify an impressively large number of species. Natalino, for instance, recorded in his schoolbook the shape and the colour of many different kinds of birds' eggs. His journal for 1937 records how he hears the *chiù* (cuckoo), then he discovers a nest of a finch with three eggs 'white, with blue drops'. Then, in the trunk of an olive tree, he finds a nest 'on the way', with the small birds already born, of a *rampichino* (tree-creeper). In the nest of a *santasecia* he finds six eggs 'blue and a little dropped with brown'. In April Natalino is ill in bed, yet he observes from the window the swallows making their nests. In the following days he discovers the nest of a tit in the wall, then that of a *codibugnolo* (long-tailed tit) and of a *calenzolo* (greenfinch).

Natalino described with equal interest and precision the birds shot with a gun by his older brother Giulio. Giulio made his own cartridges for shooting at home, filling the shells with powder and small, home-made, balls of leather. A skilled hunter, he was even able to kill enough birds so that he could sell them. Once Natalino recorded the remark of one of his friends: 'This is not a house, it's a butcher's.'

It was not only the boys but also the girls who behaved in this same ruthless way. Fernanda Caroti, on a hot day of June, was sent to drive away the sparrows from the cornfields. She knocked down a bird 'with a stick. I seized it, and told it, "If you had behaved better, and if you were less spiteful, I would have spared your life. Instead, you'll pay for it." And I knocked its head on a tree and the sparrow died.' A very drastic punishment indeed! Yet this kind of behaviour was not exceptional. When the new vice-bailiff told her to kill a young hawk, Fernanda went to the nest and 'I crushed its brain with a stone, for I was not able to do it with my fingers, as soon as I touched the brain it died.' Immediately after having buried the bird, she ran to water her flowers.

The children described very similar attitudes to wild animals. Often the children's cruelty had no justification at all: wild animals were killed just for fun. Thus Fernanda, a sensitive girl who loved flowers, recorded in her diaries, quite indifferently, pulling off flies' wings, or frogs' skin. Similarly Natalino impassively recorded the desperate jumps of the porcupines that he had thrown alive into a cauldron full of boiling water: 'they started to screech and to jump in the cauldron, then I got a razor, I removed all the quills and I cooked them'. Snakes, lizards, frogs and toads were regularly killed, often after having been tortured. Hundreds of cicadas and hornets were captured and then pelted with stones. A similar destiny was the fate of the small mice who lived in the straw-stack, or of the weasel's puppies. These last were thrown into the chicken house, cast into the role of miniature gladiators destined to be torn to pieces and devoured by the 'chicken-lions', to the immense amusement of the young peasants looking on.

The snake was undoubtedly the wild animal which aroused the strongest storm of hatred among the peasants. When a snake was discovered, there was a general competition in order to kill it as soon as possible. The animal itself was very often portrayed in a mythical way. Marcella Pampaloni wrote that a big snake was swimming in a puddle: 'it was swimming around, singing'. She ran to tell the news to her father who, exhausted after a heavy working day, was sleeping in his bed. As soon as his daughter told him that there was a snake in the field, the peasant jumped from the bed half-asleep, ran to the puddle, and bit the snake to death with an amazing number of bites.

Children and wild plants

The relationship between children and wild plants is also interesting to consider. In the San Gersolé archive there are plenty of coloured drawings of the leaves and fruit of wild plants. Their execution is aesthetically perfect, with something of a Japanese touch. In the 1940s and 1950s Maria Maltoni published two books based on the San Gersolé archives, making use of both the children's journals and their drawings. The drawings attracted the interest of important Italian intellectuals, such as Emilio Cecchi and Italo Calvino. They stressed how perfect the hands of these young peasant painters were: and they went on to make much wider generalizations about peasant culture how it was a culture closer to nature, and it was this closeness which guided the hands of the semi-literate painters in such a wonderful way.[3]

But the reality was very different. In fact the drawings were far from spontaneous. The children often wrote in their journals, or speak now during my oral history interviews, of how their drawings were the result of very severe amd controlling methods used by Maria Maltoni. She could be highly authoritarian with her pupils if they didn't manage to paint 'well'. Marcella Pampaloni even remembers Maria biting her, when she judged a painting as 'no good'. Fernanda Caroti also remembers how 'evil' she was on a similar occasion.

Nor did their beautiful paintings of leaves, fruits and flowers really imply that these young peasants felt a love for trees or other wild plants. Certainly the wood near their family holding was important to them as a place for fun and for hunting. Yet the very name of the wood in peasant dialect implies their instrumental way of looking at it: *ragnaia* from *ragno*, spider, the name of the net used for catching birds. So the wood was primarily a hunting place for both children and adults. It was also somewhere to gather wood for the fire, or for repairing agricultural implements, or for putting in a new beam in the house. In short, trees were seen as the living containers of many potential uses.

In the highly cultivated sharecropping region of Tuscany, woods were the scanty survivors from the huge original forests which had been destroyed centuries before. This destruction went on until the mid-twentieth century: woods were still being cleared to make way for new fields up to the very end of the sharecropping system in the 1950s and 1960s. Woods were also cut by charcoal burners, who had developed a technique of making charcoal by putting wood together and burning it under a covering of earth. Both peasants and charcoal burners certainly had developed very striking skills of many different kinds, which enabled them to use and modify the natural world which surrounded them in a highly effective way. But there are no indications that they could be romantic towards nature itself, nor any sign of a contemplative attitude towards forests.

While in the sharecropping region of northern and central Tuscany most of the forest had been felled, the story of southern Tuscany was different. Here a very large wild forest area had survived in the coastal Maremma up to the beginning of the twentieth century, principally because the malaria mosquito made the land inhospitable. Yet most of these forests were also destroyed by the 1930s with the draining and settling of the coastal marshes. Some decades earlier the Marquis Eugenio Niccolini, an unusual man who combined being landlord, hunter and writer, on one of his hunting days was standing on a small hill contemplating the wonderful landscape of the savage Maremma. 'What a beauty, isn't it?', he asked a peasant who was helping him. 'Yes, sir. Imagine how much corn we could cultivate there, once the forest was cut! Wonderful, indeed.'[4]

Conclusion

Tuscan peasants lived in a world in which most of the material contents of life were taken from the natural world which immediately surrounded them. They built their houses from stones collected from the fields; bricks and tiles were handmade from clay dug from nearby fields and fired in a kiln which often belonged to their own *fattoria*; even the lime was produced locally; and the beams and rafters for the roofs were felled from the surrounding woodlands. They were also almost self-sufficient in their diet. They made bread, which was their staple, from their own corn, oil and wine came from their plot, and they grew vegetables in their large gardens. They also bred their own rabbits and chickens, and found additional proteins through hunting game. Altogether they needed very little from the market: some pasta or rice, clothes, and best shoes (clogs were home made). The old saying, 'peasants only bought salt', is close to the truth.

Yet these Tuscan peasants, so 'natural' in the sources of their material lives, behaved towards that nature with an instrumental, ruthless greed. Paradoxically, they robbed it – almost like wild animals – precisely because their moral world was entirely focused on other human beings. They did not understand nature either morally or aesthetically.

At this time their main ambition was in fact to become less 'natural' and more 'human' through entering more fully into the market. And this was the change which was happening while the school journals were being written. After the Second World War young women wanted to marry a citizen, rather than a peasant; they wanted to live in a village or a town rather than on a farm; and they wanted to be able to buy and consume. Young peasant men left the countryside in large numbers, flocking to the towns and taking factory jobs, for they had to make these choices if they were to keep the chance of marrying their peasant girlfriends.

It was the next generation whose attitudes to nature changed profoundly. For some of the children and nephews of these peasant migrants to the towns came back to the countryside. They were returning to their own roots: yet they re-explored them with the perspectives which they had learnt from urban life. They bought a handsome but awkward old peasant farmhouse, restored it, and lived there with an altogether new appreciation of the fascinating beauties of nature. Some of them joined campaigning groups like the Worldwide Fund for Nature. They began to protest against the massive hunting of the small birds, whose life cycles they had watched in television documentaries. And they could be deeply shocked by the ruthless destruction of a single, magnificent oak.

Notes

1 The romanticization of the Tuscan peasantry can be found especially in many books and articles written by local historians and amateur anthropologists. Academic historians and anthropologists have generally confined their interest to Tuscan peasant struggles and, since the 1980s, researching on the sharecropping peasants has become much less fashionable.

For a good bibliography on debates about the sharecropping system, see Biagioli (1989). Perhaps the most important insights can be found in Clemente *et al.* (1980), especially in his own article, 'I "selvaggi" della campagna toscana: note sulla identità mezzadrile nell'ottocento e oltre'. Also of particular importance are the proceedings of a conference held in Siena (Clemente 1988).

2 I have also written on San Gersolé and Tuscan sharecropping families in Contini and Ravenni (1988). I began interviewing the former young peasants of San Gersolé, now grandparents, in 1986, at first with just a tape recorder and later with video. I chose those who had attended Maria Maltoni's school and had left some pages of writing in the archive. I wanted to expand the information I had already found in the old school books and, at the same time, to compare the new oral records with the old written ones. I was surprised to find how well the two matched together. A first book based on that research will be published soon. I am still working on much more extended research based on records of peasants in that area going back to the seventeenth century.

3 Some of the San Gersolé journals were first published in the Fascist period (Bettini 1940). Others were subsequently published by Maria Maltoni (1949, 1965) herself. Italo Calvino wrote the preface to the 1965 collection.

In 1985 an exhibition of the drawings and a conference were organized by the Comune of Impruneta (which includes San Gersolé), together with the University of Florence, the Tuscan Region, the Province of Florence and the Associazione Communale n.10, Area Fiorentina. The catalogue *San Gersolé, quaderni e disegni 1930–50* (1965) includes an interesting introduction by Anna Scattigno.

4 I was told this story by Francesco Giuntini, who is a living memoir of the Florentine aristocracy.

Select bibliography

Bettini, F., *La scuola di San Gersolé* (Brescia, 1940).

Biagioli, G., 'Le métayage en Italie centrale: un système agraire à l'épreuve de l'histoire et de l'historiographie', *Bulletin du Centre d'histoire économique et sociale de la Région Lyonnaise*, 3–4 (1989).

Clemente, P. (ed.), *Il mondo a metà: sondaggi antropologici sulla mezzadria classica* (Bologna, 1988).

Clemente, P., Coppi, M., Fineschi, G., Fresta, M. and Pietrelli, V. (eds), *Mezzadri, letterati e padroni* (Palermo, 1980).

Contini, G. and Ravenni, G. B., 'Giovani, scolarizzazione e crisis della mezzadria: San Gersolé (1920–1950). La storia della famiglie attraverso i diari scolastici e le fonti orali', in Clemente (1988).

Maltoni, M. (ed.), *I diari di San Gersolé* (Florence, 1949).

Maltoni, M., with Venturi, G. (eds), *I quaderni di San Gersolé* (Turin, 1965).

Scattigno, A., '"La leggenda dei tempi antichi". I disegni e diari di San Gersolé nella stampa italiana, dal 1940 alla prima metà degli anni sessanta', introduction to exhibition catalogue, *San Gersolé, quaderni e disegni 1930–1950* (Florence, 1965).

3 Narrating nature

Perceptions of the environment and attitudes towards it in life stories

Daniela Koleva

Over the last few years a new avenue seems to have been opened up in the application of the life history method. In addition to its traditional major fields of interest such as work and family relationships, national socialism and feminist movements, it has been increasingly applied to the study of the recent Communist past of Eastern and Central Europe. The focus of interest is usually the way in which the Communist system moulded everyday consciousness, as well as the various forms of resistance, behind the formal facade of real socialism.

Our own research project[1] was similar: we did not originally aim to include the theme of the environment, or to account for the formation and development of environmental consciousness. However, we became increasingly aware not only that our actions obviously have to take place in certain environments, but also that our memories are anchored to these environments as well. People's autobiographical narratives normally include a 'backdrop' of some environment, which seems inextricable from what happens. Particular aspects of the life story are tied to particular settings of the environment. Changing environments correspond to changes in a life; stable environments contribute to stable identities; and landscapes, once perceived, keep reappearing in memory, bringing back once again past feelings and moods. Thus environment seems to acquire the specific biographical dimension which has only recently attracted the attention of researchers.[2]

Environment appears on two levels in autobiographical narratives. First, as an implicit notion of the place or space[3] where the events take place, where people move, where they work and interact with each other. Second, more explicitly, as memories of particular places and landscapes linked to particular bits of the story and in the form of perceptions and ideas of 'nature' as something separate and even counterposed to everyday life. On this second level 'nature' was present in twenty-four of the ninety autobiographical interviews reviewed in the course of the work on this chapter.

The constant reappearance of various aspects of environment in autobiographical narratives without the asking of specific questions is evidence of how deep is our consciousness of our environment, how we project our own selves outward on to it and how closely it is connected with the images of ourselves which we create when thinking of our lives. It seems that settings and their 'mood' (*Stimmung*, as Simmel termed it) have a deep impact on our perception and memories.

To give just one example, here is how Sarah Iakov,[4] born 1923, remembers the circumstances of her joining the communist youth organization in 1941, during her last year at high school. This event turned out to be crucial for her later life. It happened during an excursion to the monastery of St Kirik on Rodopi mountain: 'a nice day in February – almost no snow. We could even sit on the ground. There were already snowdrops under the bushes . . . Small snowdrops. We picked snowdrops. We saw the monastery. There was a big, a huge tree in front of it.' The mood of the coming spring corresponds to Sarah's optimistic expectations. After this beginning the story follows of what it was like to be a Jew in Bulgaria during the war, of her underground activities and of the year she spent in prison (the latter comprises about a quarter of the whole narrative). Sarah's story ends on 9 September 1944 (the day of the communist *coup d'état*), when she meets her friends and comrades again, with a feeling that her expectations have been fulfilled.

Inevitably, a life history narrative itself occurs in a specific 'place', has a specific location. This place straddles the border between public and private space. By the very act of telling one's life, the 'private' person becomes a 'public' one. There is something paradoxical in the situation of a person narrating a life story; there is a tension inherent in this situation between the publicity of the very act of narration and the privacy of the content narrated. The autobiographical concentration upon oneself assumes the character of specific publicity. Thus the process of narration turns out to be a process of constructing, testing and proving one's identity. In a way, all persons perpetually narrate themselves, or at least perpetually conceive or conceptualize themselves. That is why M. M. Bakhtin's distinction between the author of an autobiography and its protagonist makes sense in our case as well.[5] They coincide 'in real life' but in the life story the question, 'Who am I?' is transcribed into, 'How do I represent myself?' This implies a distance, sometimes even a tension, between the self as author and the self as protagonist of the autobiography. The 'constructed' character of the autobiographical narrative makes it possible to think of it as a piece of literature and to borrow some ideas from literary analysis.

Thus, on the implicit level the idea of space or place contributes significantly to the structuring of the narrative. Certainly, in a life story, just like in any story, time is of definite importance. Events are arranged in their time sequence. But their sequence alone does not make a story out of them. The narrator needs to see and bring to the fore some sense, some purpose of that sequence, or what seems to be a causal connection at least. Thus space (in stories of migration, which form about 70 per cent of our sample) or place (in stories of continuous residence) quite often contribute to the 'plot' of the life story and are essential constituents of evoking and structuring the narrative.

This function makes it possible to try to make sense of the narrative as 'chronotopos'. 'Chronotopos' ('time-space') is a category proposed by M. M. Bakhtin in his study of the novel to express the unity of space and time and its function as an organizing centre of the narrative. According to Bakhtin, the chronotopos defines the aesthetic unity of the literary work and its genre. All temporal and spatial relations in art and literature are inseparable from each other.[6] Time is solidified, compressed and aesthetically visible (tangible) while space is intensified, involved in the flux of time.[7]

The link between time and space is an organic one. It is expressed in the notion of 'place'. The place in life stories is not just a setting for the events of the story and not just a geographical location. It is also a period of one's life burdened with attitudes and emotions. Thus, Elena Manolova[8] refers to the period of her life between 1959 and 1967 by the name of the town (Vidin) where she used to live. Her answers to the 'when'-questions concerning that period provide a geographical rather than a temporal specification: 'in Vidin'. The change of place is often a marker of a new episode in the life story. Therefore it is given special attention, accounted for and explained – for the sake of the unity of the narrative.

Thus, the logic of a biography implies a spatial dimension as well as a temporal one. Both types of the autobiographies with which I am concerned here – of migrants (most of them from rural to urban regions) and of non-migrants – are built on a topographic base. Space and place have leading functions in the structuring of the narrative. The unity of place in the life stories of rural inhabitants blurs the time-borders between generations and between the episodes of the life of a single individual, imports stability and determines the cyclical character of a person's narrative. Changing places in the life-span of a migrant marks its phases and forms its evolutionary trajectory.

In what follows I examine some of the perceptions of nature and the attitudes which these stories invoke explicitly. 'Nature' is a concept as abstract as 'culture' or 'society'. But in the life stories it is usually present as specific landscapes, places or specific objects – trees, watersprings or crossroads, 'privileged' by attributing to them some symbolic meaning or just by using them as spatial markers. On the other hand, what is being told about them is a complex amalgam of personal memories, specialized knowledge and culturally mediated attitudes. The perceptions of 'nature' can be summarized roughly into two types: everyday ('prosaic' or 'nature-near') and aesthetic ('romantic' or 'nature-distant'). They do not coincide with the dichotomy between rural and urban, nor with that between traditional and modern: such dichotomies do not capture the most important processes that have exerted powerful influences upon the life paths of our informants. The key process has been the so-called 'rurbanization' in its two aspects: on the one hand, the spreading of urban ways in non-urban regions as a result of industrialization and the development of communications, and, on the other hand, the grouping together of former rural inhabitants in urban and industrial zones. These processes were ideologically fostered as well. Joining the industrial proletariat, which was the 'leading social power', was a matter of improving one's social position and therefore implied spatial as well as social mobility.

Thus, distinguishing between prosaic and romantic types of perception is not an attempt at drawing another dichotomy, but rather an attempt to demarcate between two sets of questions – how we imagine the world and how we are engaged in it. A third type, now perhaps still in the making and only sporadically emerging in life histories is the engaged, or ecologically conscious perception. Let us look at them in turn.

Prosaic attitudes are most often found among people whose everyday activities are determined by their direct relation to their environment. Thus they do not perceive it as something separate and in opposition to the world of their culture, that is, of their action. They consider its various aspects unquestionably 'out there' as given, and thus defining their own world and activity. Rivers, trees, hills are not just 'part of the scenery' but concrete realities to be dealt with. Time is measured by sunrise and sunset, by the blossoming of the trees and the first snow.[9]

Ivan Borissov,[10] born in 1915 in a village near the Danube, still lives there. He focuses on the enduring, recurrent and collective aspects of his remembered past ('we used to' is the major narrative form), rather than on unique events that shaped the direction of his individual life. He appears rather as a witness to a way of life, than as the protagonist of his own life story. He focuses mostly upon the taken-for-granted, repetitive and habitual character of everyday life, giving plenty of detail about the technologies of making bricks, cutting wood with an axe, 'because there were no belt-saw and buzz-saw at that time', and so on. Thus, in a way, the limits of linear, biographical time have been transcended by his focusing on what is enduring and recurrent, or what is paradigmatic about his remembered past. 'My parents were farmers . . . I stayed seven years at school and became a farmer too.' 'Farmers' work is not easy . . . Well, it was hard, but we were used to that labour.' Through forging links between his individual life and that of the community – the village, the army regiment, the family and relatives – he seems to derive a sense of 'enlarged' and 'solidified' (to use Bakhtin's terms) life course.

The chronotopos of this life story retains something of the unbroken unity of folklore – the organic relation between time and space as *that particular* landscape into which almost all the events are knitted together. Everyday activities are inseparable from that particular spatial segment where predecessors used to live and where descendants keep returning. The unity of place fills in the gap in time between generations as well as that between childhood and old age (the same trees, same river, same hills). It makes all temporal borders fluid and rhythmical. The succession of generations in that local, spatially limited world appears temporally unlimited. Cycles of work and leisure, of growing and ageing, of birth and death, involve the basic realities of life. There is a harmony, a common rhythm of human life and nature. Ivan describes his everyday life following the yearly cycle of the agricultural work: ploughing, harvesting, gathering the maize, cutting wood in winter. Then he focuses upon the daily cycle: 'When we gathered the maize, we used to get up at three o'clock in the morning. We used to go to the cornfield with the ox-cart and . . . we used to cut and tie the cornstalks . . . and by daylight we used to have them already piled into hoods [cone-shaped piles].' More than once it is stressed that all necessities, except salt and gas, were self-produced – food, wine, soap, clothes and shoes. Thus, consumption appears significantly bound to production: what one eats and drinks and wears is the direct result of one's work. 'Earlier, we didn't buy sunflower oil even. We used to plant the sunflowers, gather them, take them to the oil-factory and have in return twice refined sunflower oil. It was white like . . . like sauerkraut juice. Not yellow. White. Twice refined.'

Through their direct link with nature, even food and its consumption acquired a special symbolic meaning, stressing their essentiality for life. This is quite conspicuous in a number of stories. To give another example, Maria Taneva,[11] who lived with her parents in the outskirts of Haskovo, a southern town, tells how in her childhood they used to 'keep the pot boiling'. They were refugees from East Thracia after the First World War and had to 'begin from nothing': her mother grew tobacco and her father was a blacksmith. 'I remember, we felt obliged to help with the housework as much as we could, because our parents worked all day long. We had no grannies to take care of us, to look after us.' The children were in charge of buying the bread for dinner. She gives plenty of detail about how bread used to be made; how children used to play in front of the bakery while waiting for fresh bread (the word 'bread' occurs six times in a paragraph of twelve lines). 'Sometimes we waited three batches [in order to get fresh bread] – they worked the dough up, shoved the bread in the oven; then waited 45 minutes until it was ready . . . Especially during the summer holidays we were always at the bakery. No one had to ask us to go and buy bread.'

In stories of this type, human life and human activity appear as a part of the eternal 'natural' order, and acquire their own legitimacy through it. They seem to derive their stability from nature, rather than counteract it. The latter is considered given and self-sufficient in its constant regeneration. Space and time are structured by the rhythm of human activities, which are, for their part, orientated towards the cyclical rhythm of natural phenomena – the cycles of days and seasons, of growth and fruition, etc. On the other hand, the products of human activity are linked to spatial conditions and follow the specificity of the landscape. A village is built by a river. The houses face south and back upon a hill. Roads follow the topography of the place as a matter of course. This is the result not only of technical necessity, that is of a technical incapability of coping with the ecological setting, but of a certain virtue, if one can put it thus, of keeping in tune with natural forms. The earth was what gave one stability ('roots') and self-confidence. Human products were not regarded as so easily detachable from the conditions that made them possible. Therefore nature was not perceived as beautiful scenery, with a mood and individuality of its own, but as particular things to be handled, particular settings to move and work in. Community and human life are seen as part of the cycles of nature. The individual has grown older; the village is dwindling; there is less water in the river and a creek has disappeared. The children have gone to the town and the school has been closed down. That is the natural order of things. The changes in the immediate environment, natural and social, are in accord with one's own growing old. In this case the 'loyalty to place' is crucial for the making of one's identity.

With the radical and painful changes in rural life brought about by collectivization after 1944, such men and women felt no longer 'attuned' to nature. Ivan Borissov perceives this situation as the disintegration of human life and nature. One of his stories is about how in the first year after the establishing of the collective farm the villagers were made to plough a field after sunset and overstrain the animals. 'The first year, the autumn of 1950, we ploughed with the oxen in order to crop the fields [of wheat]. And there was a man . . . Tzvetko Bonev, came from Vratza [the regional centre]. He made us plough in the dark. And people told him: "Hey, an ox fell there in the furrow." And he said: "Never mind, throw it aside and go on." And some people ran away, they hardly found their cart in the dark to get home. Tzvetko Bonev was that sort of man . . . To plough in the dark.' 'To plough in the dark' is repeated and highlighted by his intonation as something unnatural, violating the harmony between man and nature.

Perception of nature, however, is not only a premise for the relationship to it, but a result of that relationship as well. Quite different is the attitude of those informants who were born between 1920 and 1940, and whose active life coincided with the years of rapid modernization and urbanization. They form the majority of our respondents. A number of them have actually brought about some of the major changes in their environments. Stoyan Stoev,[12] born in 1931 in a small town in the north-east part of the country, graduated in chemistry and migrated to a larger town, where he worked at one of the first big pharmaceutical plants in Bulgaria. He was among the first experts who started the production of antibiotics. Later he moved with his family to the regional centre, where they set up the production of yeast, and eventually they settled in the capital: his wife as a researcher in biochemistry and he himself as an expert at a central institution for control and standardization. He is conscious of his role as a witness and participant in important events. The people he mentions are introduced by their profession, education and their migrant itinerary as the most important identifiers: 'I have a sister, who is four years older than me. She graduated in medicine and worked in the town of Toulboukhin, which is now Dobrich. Then she married and moved with her family to Varna.' He has an acute sense of the changes and of his own personal participation in them. His own life seems absorbed by the events of which he speaks. The self is a contraction of the time elapsed and of the space conquered – geographical as well as social. There is a conspicuous overlap between biographical time and historical time. The plot around which the narrative is organized is provided by 'big' history. Its temporal dimension is, however, reduced to the spatial mapping of his life's career. Every change of place marks the beginning of a new episode. The story of Vera Mineva[13] is quite similar. She was born in 1935 in a village and stayed there till her marriage. Her first contact with the 'outer world' was when German soldiers came to stay in her village during the war and one of them 'gave us children a ride on his motorcycle'. Her horizons broadened further when, as a youngster, she took part in amateur performances and travelled around the district. 'We, the young, took part in the collectivization as amateur performers; we travelled around the neighbouring villages to persuade people to join the co-operative farms. Then we gave performances.' She was pleased to be part of that team of enthusiastic young people who popularized a new way of life. Later she married a man who worked with a prospecting drilling team and thus her nomadic life began. Her story then looks like a map of the places they conquered – 'the north of Bulgaria through and through', 'always in temporary lodgings, always staying with strange people'. Traditional local identity had disappeared and was reconstituted around a new focus.

Nikola Markov[14] was born in 1934, and was a construction engineer who took part in the draining of large agricultural areas along the Danube, and in the construction of the infrastructure of the north-west part of the country. He remembers the feeling of self-confidence about his work. He thinks of the environment as changed landscapes set around what once used to be building sites – something which is a result of human activity as well, and to which he himself has contributed. He describes extensively and with abundant topographical and technological details how living conditions were changed and the landscape consequently altered. His explanations follow a cause-and-effect pattern: 'We planted those trees there to prevent erosion of the banks of the stream . . . The creek disappeared because of the wood. Those hills were bare – shrubs and stones only. Every heavy rain produced rivers of mud. We reforested them, so that the trees could hold the soil. But the trees have deep roots and need more water. So they get the water from the deeper layers of the earth before it comes to the surface.'

However, this is not a glorious story of victory over nature. With a nice sense of humour he tells of the considerations which determined the present route of the road: pressing deadlines, bad weather, a conflict with the town mayor. 'You cannot imagine how many kilometres I've driven along the country roads here . . . in heavy, sticky mud . . . day and night, no weekends, in summer and winter . . . I didn't see my family for weeks. I must have been a fool to do it.' There is a tinge of bitterness about 'Communist construction' and the illusions he used to have at that time.

On the other hand, both Stoyan Stoev and Nikola Markov claim today's landscapes as partly something of their own making and thus locate themselves in the changing world. In this group in general there is an evident consciousness of the dependence of nature on human intervention, of today's nature as a human product. Thus, Boyan Lilov[15], when spending his holidays by the sea, admired not only the sun and the sand, but also the construction of new resorts such as Slanchev Briag and Zlatni Piasatsi, identifying them with the youth of his generation. Miriana Atanassova[16] mentions 'nature' and 'spending free time close to nature' only in respect of her university years (1964–9) and the years of her administrative work in a regional centre (since 1989), but not in respect to the period in between, when she was working as an agriculturalist.

Romantic attitudes and the perception of nature seem, by contrast, to be much more typical of the everyday culture of cities. The schemata of weekend family picnics ('going out', 'close to nature') persist in the narratives of townspeople. Picnics are invariably considered a respectable and positive way to spend one's weekend. Nature is represented as 'pure', 'beautiful', 'fresh', 'healthy', 'beneficial' and friendly to people. 'Everything was natural, no fertilizers . . . nothing. Never again did I eat such sweet melons after my childhood years.'[17] It seems that this type of perception has settled as a pattern of everyday culture as a result not only of the way of life in cities, but of the acquaintance with and acceptance of the (primarily western) tradition of the aestheticization of nature in literature, music and art as rooted in Romanticism.[18] Nature, or parts of it, are reified and presented as symbols of omnipresence, continuity and eternal harmony.

Attitudes towards this idealized nature are frequently very personal and emotional. Thus, Dora Dineva,[19] a descendant of mountain-dwellers who spent all her life in the city, tells of her summer vacations: 'There were no motor campsites at that time. You like the place and you just stay there . . . I've always preferred the southern sea [the southern Black Sea coast of Bulgaria] because of its friendliness, because of the sloping beach and the fine sand . . . I've always adored the sea, I even remember being torn between my love of the sea and my devotion to the mountains . . . and if I had to choose between the sea and the mountains, I used to opt for the mountains because of this variability, this upward motion; because you had to struggle up towards a summit or a cottage and bend beneath your rucksack . . . But the sea, the sea is a world that never goes away. I could never understand it . . . That's how I love the mountains and the sea.' In a number of cases nature comes to the fore explicitly when it comes to leisure – weekends and summer holidays, very often – with friends and family. As Maria Taneva, a town resident and an active member of a tourist club since 1962, put it: 'What we experienced in the mountains has brought us closer and we have been friends for 30 years now . . . I don't see how young people can manage it like that. It was really glorious and people were somehow more good-natured.' Personification and projection on to nature of human emotions and human desire for peace and support mark the romantic attitude as well.[20]

There is no doubt that the common career pattern of the 1950s and the 1960s, migration to urban jobs and contexts, has significantly contributed to the aestheticization and individualization of nature in the romantic attitude – the migration from the country, where one has spent one's childhood and probably youth, to urban and industrial areas. A deeply emotional attitude towards one's own childhood and youth is interwoven with a nostalgic longing for that particular place where those years were spent. Thus Dora Dineva said of her roots: 'My father . . . there were three sons, just like in fairy tales. Mother a widow. Zheravna is a mountain village . . . I adore it and I think, I can smell Zheravna even with my eyes closed. The scent of acacia in spring, of boxshrub . . . The purity, the murmuring of the water . . . and those fountains – each with water of distinctive taste and each with its own history . . . The village is situated on a southern slope and there are small woods up the hill. Water will always run down from there. Every spring and every fountain has its own history.'

Dona Zareva[21] was born in 1925 in a beautiful village on a river-bank, near the medieval capital of Bulgaria. She moved to Sofia after her marriage and worked as a school teacher until her retirement. The memories of her childhood and school years and of her native village comprise more than half of her story. She says of her childhood that it was 'a fairy tale' and goes on to talk about life 'close to nature': simple but healthy home-produced food, vegetables, honey and thick buffalo milk; planning the daily and yearly activities according to the rhythm of nature; bathing in the river, playing outdoors and living together, in harmony with other people. 'There you have – healthy food. Healthy life. People get up early, work, look after the cattle. Early in the morning, when it is still fresh and cool, they go to the fields and when the sun begins to scorch, they hide in the shadow and rest on the warm ground. They take a bite and work again in the afternoon. That sort of regime they had and that's why they were healthy.'

It is above all in the stories of those individuals who spent their early years in the country, but, later on, lived away from it ('far from nature'), that we found images such as of 'mushroom and herb gathering', the 'family picnic', 'virgin forest', or of 'childhood in a small mountain village', 'the hardships and joys of agricultural labour' or 'living close to nature'. Perceptions of nature, patterned in this way, are quite common among migrants and town residents. Stereotypical as they may seem, they appear to play an important role as anchoring points in the self-reflection of their lives. The past here somehow becomes more real, more true to life than the present: 'to look at the [river of] Maritsa and know how deep it used to be, not just a stream as it is now, how many fish there used to be, how it used to freeze in winter . . . There were real winters at that time and everything was normal, I mean, winter was winter and summer was summer . . . and now everything's confused because man has decided to remake even nature itself.'

The romantic attitude can probably be seen in relation to H. Lübbe's idea of 'musealizing' (*Musealisierung*): a process which takes place in contemporary culture in relation to history as well as in relation to nature. With regard to history, 'musealizing' means the rapid growth of the number of 'relics', 'museum items'; it means the accumulation of our past into our present as a result of the rapid progress brought about by modernity. Pre-modern societies used to have shrines, with the aura of authenticity and rootedness in place. Modern societies have museums as a result of a deliberate effort to evoke a sense of the past, and to cultivate a sense of place. This type of historical consciousness has been expanded into nature as well. During the nineteenth century, alongside the art museums the first museums for natural history were established. Today's environmental consciousness is, according to Lübbe, a result and part of the 'musealizing' of nature,[22] of the effort to preserve (or rather, to construct) an authenticity of nature in the same way in which local traditions have been constructed and cultural heritage invented. In the realm of culture, modernity does not actually abolish the non-modern world but relativizes it by preserving and reconstituting it as a museum item. In the realm of nature, it acts in the same way, constructing the historicity of nature and inventing nature as a counterpole to civilization, to technologies, to law and custom, to culture, or even to man.

The romantic attitude, combining aestheticization and nostalgia, persists right through a number of narratives, fusing journalistic clichés ('destroyed nature', 'dying nature') with very personal and deeply emotional attitudes, tied to circumstances and events of a person's own life – most often childhood and youth. Maria Taneva, who used to spend the weekends in the mountains with her family for years, feels now that it was not only the mountain itself, but the presence of her children as well, that cheered her up: 'Now that the children have grown up and left home, we keep on going to the mountains, but it is not the same. As we start to climb and they are not around, I think of them, I miss them and it is somehow boring.' The charm of nature turns out to have depended on the family atmosphere which is no longer the same. So the mountain has lost much of its attractiveness.

An ecologically conscious attitude towards environment is much rarer in these memories. It could be derived from two different sources.[23] Sometimes we find in opposition to the idea of nature as eternal order and regularity the notion of a lost harmony, the image of a threatened nature: 'everything's confused because man has decided to remake even nature itself', as Dora Dineva put it. Polluted nature, in its turn, threatens the health of people. Dona Zareva emphasizes that in her childhood people used to be much healthier than now because they consumed natural foods 'produced with natural compost, no nitrates, no pesticides' and stored without chemical preservatives. That is why, according to her, 'Diseases were rare . . . There were no [they were unnecessary] antibiotics and special medicines.'

However, it seems that it is the type of knowledge usually called 'expert' that most often leads to ecological consciousness. Ecological consciousness is 'remote from common sense',[24] that is, far from the world of everyday attitudes. According to H. Lübbe, this results rather from conflicts between instrumental and practical reasoning.[25] Day-to-day interaction with the environment has an instrumental character and does not lead to perceiving it as nature, that is, to a more or less detached perception. The functional rationality of everyday relationships with 'environmental resources' turns out to be anarchistic, in as far as the environment itself is concerned. It can, however, lead to the view of nature as 'socialized' (Giddens), 'domesticated' (Elias), 'anthropogenic' (Böhme), and to the idea of human responsibility for it. This idea exists in some of the life stories as the present position of the narrator, which is not discussed in detail. Thus, Nikola Markov incidentally mentioned that the construction of dam lakes and hydrosystems sometimes put plenty of fertile land to waste, as did the construction of a new industrial town, Dimitrovgrad, the 'valley of the Big Chemistry', as they proudly called it. Kera Stankova, who spent her childhood in a small mountain town, now the centre of a mining region, also mentioned the damage brought about by mining, but at the same time remembered how with the coming of the first geologists 'a whole new world opened before us children'.

It is only natural that our informants, now in their sixties and seventies, focus in their life stories on a period when ecological damage was not as conspicuous as it is now and ecological problems were hardly ever discussed. Most of the stories they consider worth telling end well before the 1980s and that is why even such an ecological disaster as Chernobyl is never mentioned.

As we see from the above discussion, environment is present on several levels in the autobiographical narratives: as the spatial dimension of the life stories, as the particular settings in which people move and work, as immediate facts to be dealt with in everyday activities and (partly at least) as a result of these activities, as an alternative to urban culture and a symbol and source of health, peace and harmony. The oscillation of attitudes between the poles of dissociation, estrangement from nature (as in the oppositions 'man–nature', 'culture–nature') and engagement with it (as in the idea of 'civilized nature') reiterates the ontological paradox of human beings, who, being nature themselves, exist apart from and counterposed to what they regard as nature.

Notes

I am indebted to the anonymous referees whose comments were of enormous help in improving this text. Special thanks also to Mr C. R. Beswick for correcting my initial translation into English.

1 This chapter is based on the research project 'Experienced History' which has been under way since 1995 at the Centre for History and Theory of Culture at the St Kliment Ohridski University of Sofia, with the financial help of the Bulgarian Ministry of Education and Science. The objective of the project is to try to fill in some significant blanks in our knowledge of everyday life in Bulgaria during the last fifty years. The life history method has been employed to trace the ways individuals experience historical events.

2 See for example: A. Lehmann, 'Wald als "Lebensstichwort". Zur biographischen Bedeutung der Landschaft, des Naturerlebnisses und des Naturbewußtseins', *BIOS*.

Zeitschrift für Oral History und Biographieforschung, 2 (1996), 143–54; from a slightly different perspective: N. L. Peluso, 'Fruit Trees in an Anthropogenic Forest: Ethics of Access, Property Zones and Environmental Change in Indonesia', *Comparative Studies in Society and History*, 38(3) (1996), 510–48; M. Donovan, 'Capturing the Land: Kipsigis Narratives of Progress', *Comparative Studies in Society and History*, 38(4) (1996), 658–86.

3 For the purposes of this chapter I distinguish between space and place only in relation to the two most common types of life career: one resident, and the other including migration from the country to a city and change of occupation. This distinction could be embedded in the context of more fundamental conceptions of space and place, like A. Giddens, *The Consequences of Modernity* (Cambridge, 1990) or D. Harvey, 'From Space to Place and Back Again: Reflections on the Conditions of Postmodernity', in J. Bird (ed.), *Mapping the Futures: Local Cultures, Global Change* (London, 1993), 3–29. That, however, goes beyond the purposes of this analysis.

4 Interviewed February 1996.

5 'Author and Hero in the Aesthetic Activity', Bulgarian edition, 'Avtor i geroi v esteticheskata deinost', *Filosofia na slovesnostta* (Sofia, 1996), 168.

6 M. M. Bakhtin, 'The Forms of Time and the Chronotopos in the Novel', Bulgarian edition, 'Formite na vremeto i hronotopa v romana', *Vaprosi na literaturata i estetikata* (Sofia, 1983), 439.

7 Ibid. 272.

8 Born 1926, interviewed November 1995.

9 Kerana Stankova, interviewed April 1996.

10 Interviewed 23 October 1995.

11 Interviewed August 1996.

12 Interviewed 17 November 1995.

13 Interviewed 2 November 1996.

14 Interviewed August 1996.

15 Born 1932, interviewed November 1995.

16 Born 1945, interviewed September 1996.

17 Neda Siderova, interviewed July 1996.

18 According to Simmel, the 'sense of nature' (*Naturgefühl*) existed long before Romanticism, in primitive religion, for example. What Romanticism did develop was a sense of *individualized* nature, that is, of landscape as an entity with a mood and meaning of its own. The ability to see a landscape instead of a sum of objects is essentially an aesthetic one and its archetypic form is the landscape painting. See G. Simmel, 'Philosophie der Landschaft', in *Das Individuum und die Freiheit: Essais* (Berlin, 1984), 130–9.

19 Born 1933, interviewed September 1995.

20 Norbert Elias notes the tendency to personalization and reification in the contemporary view of nature. See N. Elias, 'Über die Natur', *Merkur*, 40 (1986), 469–81.

21 Interviewed September 1996.

22 H. Lübbe, *Fortschritts-Reaktionen: Über konservative und destruktive Modernität* (Graz, Styria, 1987), 157.

23 A third source remains completely outside the discussion, because it is not represented in our sample: the ecological movements and groups (primarily of intellectuals) which appeared in the 1980s in some of the larger towns in Bulgaria as a form of opposition to the Communist government and which during the last years transformed themselves into political organizations.

24 For 'common-sense-fern', see H. Lübbe, *Der Lebenssinn der Industriegesellschaft: Über die moralische Verfassung der wissenschaftlich-technischen Zivilisation* (Berlin, 1990), 143.

25 Ibid. 146.

Select bibliography

Bakhtin, M., *Filosofia na slovesnostta*, vol. I (Sofia, 1996).

—— *Vaprosi na literaturata i estetikata* (Sofia, 1983).

Donovan, M., 'Capturing the Land: Kipsigis Narratives of Progress', *Comparative Studies in Society and History*, 38(4) (1996), 658–86.

Elias, N., 'Über die Natur', *Merkur*, 40 (1986), 469–81.

Giddens, A., *The Consequences of Modernity* (Cambridge, 1990).

Harvey, D., 'From Space to Place and Back Again: Reflections on the Conditions of Postmodernity', in J. Bird (ed.), *Mapping the Futures: Local Cultures, Global Change* (London, 1993), 3–29.

Lehmann, A., 'Wald als "Lebensstichwort". Zur biographischen Bedeutung der Landschaft, des Naturerlebnisses und des Naturbewußtseins', *BIOS. Zeitschrift für Oral History und Biographieforschung*, 2 (1996), 143–54.

Lübbe, H., *Fortschritts-Reaktionen. Über konservative und destruktive Modernität* (Graz, 1987).

—— *Der Lebenssinn der Industriegesellschaft. Über die moralische Verfassung der wissenschaftlich-technischen Zivilisation* (Berlin, 1990).

Peluso, N. L., 'Fruit Trees in an Anthropogenic Forest: Ethics of Access, Property Zones and Environmental Change in Indonesia', *Comparative Studies in Society and History*, 38(3) (1996), 510–48.

Simmel, G., *Das Individuum und die Freiheit. Essais* (Berlin, 1984).

4 When the water comes

Memories of survival after the 1953 flood

Selma Leydesdorff

The Disaster

On the night of 31 January 1953 a massive flood struck Zeeland in the southwest part of the Netherlands. It was a flood of such epic proportions that it is known simply as the Disaster.[1] This flood killed more than 1,800 people. Vast areas of land were inundated, countless livestock were killed, and roads and houses were destroyed. The Disaster left an enormous toll of suffering in its wake. Recollections of this flood form part of the mythology of the national identity and memory that has to do with centuries of victory over the water. Indeed, memories of the event remain vivid for anyone who was alive at the time. Outside the flood region, many Dutch people recall how they collected blankets for the relief effort. Hundreds of others came to the stricken area in small boats, which unfortunately hindered the military rescue efforts. It is equally interesting that many people who were not even there recall striking images of the water.

In the aftermath of the Disaster the national reaction was one of pride. Had not the whole population of the country collected money and goods for the survivors? Had not every town and village sent its young people as volunteers? The dominant account of the tragedy turned the Disaster into a victory in the battle against the sea, as symbolized also in the construction of a mammoth system of dikes and waterways called the Delta Plan. This story suppressed loss, sadness and fear while highlighting national collective action.

If you travel to Zeeland you will find that the memory of the flood surfaces in the life story of virtually every local resident, but that each memory is different. There is a certain confusion, and the stories have mutated over time. An individual's own recollections are now intermingled with the memories of others such as relatives and neighbours. It may no longer be clear who was present during that night – in the attic, in the boat or in the terrible water. The Disaster made chaos of the world and threw people into extraordinary situations. To the survivors it was a deluge of biblical proportions, leaving nothing unaltered. They still cope today with chaos, loss, repressed memories of grief, and recurring anxiety. Their recollections have nothing to do with heroic victory over the water.

Between 1988 and 1993 over 200 interviews were conducted in the region. The commemoration of 1993 was approaching and many in Zeeland and Brabant felt that their history had not been told yet, that their memory was not included in the national version of the event. The flooded territory was a remote agricultural society of small farmers, whose very strict form of Protestantism affected every aspect of daily life there. Since that time, there have been dramatic changes in the way people live owing to rapid industrialization and exploitation of tourism. Transportation, previously mainly by boat, is now by car over the enormous dikes that connect the isles with the mainland. Where there was once silent country, in which one could hear the wind and the sea, there is now a playground of small yachts, campers and motorcycles. This dramatic change adds an additional layer of uncertainty about what happened during and after that terrible night. Was the world safer then, or is it safe now? Better then, or better now? Such questions are inseparable from memories of the past.

A survey conducted by the University of Amsterdam in 1991 showed that 86.2 per cent of the population feels protected from the sea by the network of dikes. The people from flooded areas reported feeling even safer than the rest of the Dutch population. However, people from Zeeland responded differently to the survey questions than others elsewhere who were also of an age to have experienced the Disaster.[2] It is unclear if this is the result of hearing so much about the Delta Plan; possibly the survivors want to believe what they have been told about the level of safety it provides.

What does feeling safe mean?

How one feels about immunity from natural disaster is relative. Feelings of safety are culturally determined, and our sense of danger varies over time. For example, what is happening to the natural environment was just as menacing twenty years ago as it is today, but now the public is more aware of it. The same is also true of crime. And yet the Dutch feel safe and protected from the sea, despite simultaneously being aware of the rising sea level and the melting ice cap.

Interviewing survivors of the Disaster forces one to re-examine feelings of safety. The manner in which past tragedies are remembered and re-evaluated affects how we assess the possibility of a repetition. In the case of Zeeland, the narratives of the Disaster are also influenced by the deeply religious attitudes of the population. This is especially the case because the worst hit areas were the most remote and most religiously orthodox. There, skirts and sleeves are worn long, television is often forbidden and the churches remain full. Many consider their fate to be in the hands of God, and many believe that the Disaster was simply God's punishment on them.

The American sociologist Kai T. Erikson, who has studied the destruction of an Appalachian community by flood, also examines this issue.[3] In contrast to my findings, Erikson reports that to 'blame' God by saying that the flood was His punishment was considered blasphemous by the people of Appalachia. When journalists suggested that interpretation, people would get angry with them. As Erikson states, 'No scriptwriter hired for the purpose could have composed a line better calculated to irritate the people of the hollow. In Appalachia, God is not blamed lightly for the mistakes of men. To do so is to risk something close to blasphemy, not just because it employs the name of the deity too casually, but because it seems to accuse God Himself of a terrible wrong.'[4]

Erikson quotes a local resident: 'The big shots want to call it an act of God. They have told a lie on God, and they shouldn't have done that. God didn't do this. He wouldn't do that.' In that case, the collapse of a dam is understood historically as a man-made disaster, whereas Zeeland's flood has gone down in history as a disaster without explanation.[5]

That particular dam in Appalachia collapsed suddenly owing to the miscalculations of a mining company, which built dams as a cheap way to dispose of waste. But no one there ever thought he or she might be drowned. By contrast, in Zeeland the people shared a collective memory of past disasters. Among these the Elisabeth flood of three centuries earlier was the most famous, and the last major flood had occurred in 1911. But since that time the land had never been inundated, except to prevent the allied forces from landing in the last phases of the war, in 1944. On that occasion, when the flooding was deliberate, no one was killed. Yet although by 1953 floods were deeply embedded in the collective memory, waters were not expected to rise above knee-level.

The residents of Zeeland were rightly anxious on that evening at the end of January 1953. Many of those interviewed mentioned the large groups of men who regularly checked the wharf and who had been at the harbour that night. The flood boards were in place, and the prevailing sentiment was that things would turn out all right. As one man recalled:

> People just went to bed with their doors tied shut [a common practice during heavy storms]. This was not the first raging storm that had struck – of course the people who live near the harbour were more aware of what was happening. Some stayed awake all night or woke in time. We weren't very watchful. Someone came to warn us.[6]

Another man said:

> My wife set the table for Sunday morning, and then we were all done. We went to bed. Then around two, or two fifteen, we heard fire engine sirens. That scared us. A fire during a storm like this one was not a good sign. So we got up and got dressed and went outside. We heard people running and the sound of their wooden clogs.[7]

The storm was unrelenting; men tried to hold the dikes, even standing in the water to do so. In places the water came in like an enormous wall. And afterwards the flooded land remained under water for more than a year.

Of course the Disaster was the result of extreme meteorological conditions, but years earlier official reports had warned about the inadequate height and structural unsoundness of the dikes in Zeeland. At the time of the flood most people were not aware of this, but when they remember the flood now, they know about the poor condition of the dikes in 1953. People know now that they could have blamed the government at the time, which creates a confusion in their memories. One easy solution to this confusion is to find an explanation in the revenge of God. This religious explanation, however, is contested even by some of the strictest Protestants, given the public knowledge of how much improvement the dikes needed.

Uncertainty and confusion have made people scared of a repeat disaster. This fear that adverse fate will strike again has deepened their trauma. In the villages no one speaks about this, since everyone knows what others have endured. As is normal after a massive trauma, there has been silence, anxiety and severe depression. All of these are expressed in narratives which are not about the Disaster as such, but about religious or political disagreement. They also underlie the stories that villagers tell about the bad behaviour of others.

As many studies of disasters note, depression leads to thinking that Evil will try to catch you. All of this is woven together in the story an elderly lady told:

> It was not God's punishment, that's nonsense. That's just impossible, it really is. You know, God is a spirit, but a spirit is inside people. It doesn't turn up by itself, but it did turn up then. Look around you – everything has turned out for the better. So much progress has been made. Nowadays you cannot see it [the Disaster] as an Act of God. Nowadays you can't say, Yes, I prayed for something and then it happened. We aren't little children; we can tell such tales to children, but someone with sense won't believe it. Church people don't believe it either. No question about it. I'm not really a church person, but I have been a churchwarden.

But she went on immediately to predict the return of the Evil:

> Something is coming in the near future. A great danger, a huge disaster is coming, although we might not witness it, maybe not even you . . . But if things go on like this, in this way, then the Islam is coming. And they [the Muslims] want everything. They will take your wife, and you can leave – you don't have to study, since you're a woman, men will do that for you. The danger is human lethargy.[8]

This informant sees danger all around, so that in the course of about one minute she moves from the punishment of God to the rebuilding of Zeeland, and then on to a new and possibly equal 'disaster' in the future: the arrival of immigrants and a new religion. The confusion of all these elements in her thinking itself expresses her fear of dying 'unnaturally'.

Perceptions of death

One problem with interviews about natural and environmental disaster is that death is not considered acceptable if it is not perceived as 'natural'. We fear especially anything which threatens to shorten our lives prematurely, to stop us from living until death comes in what we consider to be a normal way. This is a subjective notion of safety and lifespan, related to the modern idea that the human species can master nature. For our own peace of mind it is better not to contradict this notion of human capabilities. Although modern technology has become a new enemy, it can also still be our best ally. The philosopher Lévy Bruhl[9] argued that modern humankind assumes that death will result from natural causes. This is in contrast to what he calls 'primitive' humankind, who accepted accidental death and often assigned a mystical explanation. In those instances the questions were, why has this happened and why were those particular people the victims? The controversial book by Mary Douglas and Aron Wildawsky, *Risk and Culture*,[10] also addresses this issue. Douglas and Wildawsky suggest that fear of environmental death is 'only' a part of our culture and thus far less a real possibility than it seems now.

Death by water – by a force of nature – is not a normal death, nor is any death by environmental catastrophe. This is not acceptable in our current cultural pattern. It is seen paradoxically as preventing us from achieving a natural death, while also reminding us that nature cannot be wholly mastered. What has struck me most, not only in interviewing but also in reading about disaster and catastrophe, is that the working out of trauma cannot be done within contemporary western cultural patterns, which offer little when we try to understand death. We must fall back on older cultural perspectives within which such deaths can be explained, and which allow people to seek reasons, to be scared and to try to understand. The following may serve as an example.

For part of each year I live in the French Alps, near a mountain famous for rockfalls and stone avalanches. The valley below it is a tremendous source of legends. The greatest disaster was in the middle ages, when a huge section of rock fell off the mountain. The last small avalanche, which killed several people, was in the 1950s. An awareness of the moving mountain is very much alive in this area, and local residents display the keenest interest in knowing what is inside the mountain. This interest has little to do with the science of geology or the farmers' concerns regarding water supply (which comes from the mountain).[11] Rather, it has everything to do with the living memory of a major disaster. The danger is larger than life, even archetypal; the mountain is a vengeful entity that lives in people's prayers and in the fears of children. Every year there is a procession to its peak, and it is said that women pray for the mountain not to fall on our side, but on the other side. (Up till now the mountain has always fallen on the other side, so it is unlikely that any stone will fall on our side.)

Some of this behaviour is rational (if one believes in God, then praying is rational), some is rooted in history and some is 'superstitious'. All of this behaviour, however, is at odds with what we regard as modern rational culture. And in the end the fate of the valley is in the hands of God. I must confess that sometimes even I believe in the power of the mountain when it is wreathed in black clouds and seems so angry.

We need explanations as to why some die while others survive, and we take many different roads in search of such reasons. Within contemporary Dutch culture, the fact that the dikes gave way is regarded as an unfortunate event that must be avoided in the future. But this is not a suitable answer to the survivors, for to them loss is not rational. The reason for the deaths of so many is, in a sense, irrelevant; it is not possible to know why certain individuals perished and others did not. The point is that people feel these deaths were not normal or natural, and hence they keep questioning.

Breaking the silence through interviews

A common reaction, found in many studies, is for survivors to remain silent, because there is no accepted framework in which to talk about such traumatic events. Indeed it was only within the context of collaborating for this study that some of the people interviewed found a place for their narratives. Many told me that this was the first time they had really talked about the Disaster. Even more striking was the fact that my research gave them a structure in which they could discuss their personal experiences. One man put it this way:

> I hope the combination of all these memories will make future generations stop and think. I hope they will wonder how people could have had such horrible experiences. I hope that future generations will learn to appreciate these stories. The idea that my experiences might benefit others makes dredging up all the horrors worthwhile, that's why I did it. But I wouldn't let my experiences be turned into a newspaper article. No, I wouldn't agree to that.[12]

At the beginning of my research in Zeeland I was amazed by people's inability to speak openly and to express sorrow; there are no monuments where they can mourn together, only graveyards with individual stones. In the local press one does find regional forms of poetry expressing, often in bombastic terms, feelings that could not be expressed otherwise. At a certain point after the Disaster, words about it subsided, and survivors' memories were recounted only on special occasions. Every year around the first of February, however, articles, editorials and other writings reveal the shortcomings of the national myth by expressing an alternative to the story of courageous struggle against the sea. In local libraries in Zeeland there are bookshelves full of accounts by individuals. Many of these barely extend beyond the dikes of the narrator's island. These reports remain unquestioned, but their language is cold and impersonal.

In interviews, it was clear that the official version of the flood that prevailed bore little resemblance to the victims' own memories, and many survivors even wondered if their experiences warranted description in this study. After nearly four decades of silence, it was difficult for some to grant an interview. Only later in my research did I begin to understand that people who survive such a catastrophe don't feel that talking might be useful, or don't dare to speak. It might be better, they seem to feel, to leave their experiences frozen, isolated and untouched.[13] Communication among them is not necessary since they guess and know each other's suffering and are furthermore afraid to hear it spoken. In interviews pictures are conveyed, not the anxiety and fear that lie behind the pictures. An example is in the interview of Mrs D. She described sitting, cold and wet, on the roof of her house after morning had come, and trying to find out which houses were still standing:

> There was a dinghy with five little kids, all alone. It was enough to make a person sick. We wondered whether anyone could reach them. We looked at those kids' hands. I kept seeing that little blue mitten in my mind for months. First we had clung to the chimney on the roof and felt our clothes blowing in the wind. I felt like a flag. I watched that dinghy. After a while only one of the kids was left. Then all I saw was a little hand waving. And I thought I would be next to go.[14]

There is official mourning, and at such moments the newspapers retell the event in a rather impersonal manner. But despite this formal recognition of the event, there is no accepted way to speak about personal suffering and pain. Survivors are afraid to discuss their innermost feelings; they assume that to do so would make them seem pathetic. They also assume that the other survivors already understand: thus there is nothing to tell.

One man I interviewed was 15 years old when both his parents perished; he had been evacuated to relatives who had not been hit by the flood. He spoke of his loneliness in coping with his grief when he returned to his parents' village. Nobody wanted to speak to him, except one of his deceased father's creditors:

> Can you imagine how frustrated I was! I was only a boy, fifteen years old. You would think people would say something to you, but no one did. Yes, there was one – yes, I remember. The shoemaker. He had lent money to my father. My father had wanted to own his own house, you see, and so he had borrowed some money from him. In those days you didn't go to the bank, but you went to someone who had a lot of money. Well, that man grabbed hold of me and said, 'Your father still owes me 2,700 guilders.'[15]

It is also possible that, because of the flood, many people lost a sense of community, and it never returned. This has been a major concern for many survivors, and the enormous changes in lifestyle, geography and even something we might call modernization have left even larger scars. All this leads to more confusion in the memories surrounding the event.

Victims of environmental catastrophe are able to talk about their feelings; they simply are not likely to do so within their own communities. Stories about catastrophes are subjective, and are coloured by a shifting set of fears regarding the present. The root of this fear is the idea of an external force in nature – God, or whatever else there might be beyond human control, against which we as humans are powerless. As one woman said: 'I live here on the isle with the fear of whether it will happen again. They tell me: "It's not possible. Nothing can happen anymore." [Her voice rises] But the wind howling through the trees at night keeps me awake in the winter. My husband sleeps through it all the time. He wasn't there then.'[16]

Personal narratives

The flood in Zeeland occurred only eight years after the end of the Second World War. A mood of recovery and renewal prevailed, of combined effort to rebuild the country and to repress the memory of the war. Words such as modernization and restoration reflected the tenor of the times and the push for consistent progress. National morale was high in 1953, as was support for the Delta Plan. However, the motto 'Open your wallets to help seal the dikes' failed to meet the deeper fears and upheaval suffered by the victims. Their personal stories convey quite a different mood.

In the interviews victims told of icy water that was never far away, fear, helplessness, bone-chilling cold, a rain storm, a putrid stench, brine, mud, and finally the massive cleanup. They felt deserted by the rest of the world, as if no one knew of their fate; they feared being drowned like the rest at any moment. Then they spoke of how they survived and managed to get on with their lives. The national image of the Disaster's victims is confined to thrashing around in the water. This has tended to obscure their unselfish (and occasionally selfish) behaviour, and the subsequent constant struggle to survive and rebuild their lives.

The most heart-rending memories involved narrow escapes or near rescues. Some almost drowned but survived. Many floated for hours or even days on a piece of a house that might have capsized or sunk at any time, only to drown when trying to climb onto a dike. Others did make it to safety. But everyone faced death. During the course of the interviews it became clear how the threat from the sea has become part of the culture of the flooded areas, how this subjective feeling of being unsafe is strengthened by memory, and at the same time how it helps to structure memory. Survivors of the disaster feel that threat deeply. As one of the interviewees said:

> In a really heavy storm, I'm never at peace. Because I've learnt just how powerful the water is. If you consider that the water rises three feet above

> normal, and then you consider that the entire North Sea is pushing against that same three feet of water level – those are immense forces. What can you really do against that?[17]

Nevertheless, if asked in a formal way if they feel safe behind the dikes they will say: yes. The personal level is far away from collective memory and opinion. Whoever listens to the interviews is overwhelmed with repetitive but highly individual detail. One wonders therefore if it is possible to discover common bonds and to summarize the experience. Anecdotes, flashes of memory and endless elaborations add a scattered human touch to the teller and give the listener some keys as to how one remembers. That is precisely where the research has interest: in examining the details and asking in what context they have meaning. In the Appalachian case, Erikson also notes that 'phrases that do a particularly apt job of capturing a feeling common to many people may have circulated up and down the hollow, expressions that strike a common chord may have come to serve as a group explanation for what are otherwise individual emotions'.[18] Much more attention should be given to the loss of what he calls communality.

But repetitions and all, nothing matches the unique expression of the first-hand account in describing what happened to those who were miraculously rescued:

> We were sitting on the roof. But at a given moment, well, there *was* no roof, everything sank into the water. I landed on a part of a small shed, but not my father or brother. My father was on a roof. There were still tiles on it. And my brother landed elsewhere. And I sat on the shed, but it was too flimsy, much too light, and not suitable. Then I grabbed a piece of a beam, a good-sized piece from the shed with the collar still on it. It floated by and I could grab hold of it and crawl onto it. And so I could stay above water on the beam. My brother was farther up, he was a couple of hundred yards ahead of me, he was floating on a piece of roof, probably from a shed or a stall, I'm not sure. My father was also rather far away, more to the right, kneeling on a piece of roof with the tiles still on it. And so we drifted, in the direction of the village of Ouwerkerk.

Ouwerkerk is located on the coast, and whoever floated past the village would end up on the open sea. But at that moment this man did not realize that he might drift out to sea.

> I was moving in the direction of the hole in the dike. I didn't realize at the time that there was a breach. That hole in the dike, I didn't know a thing about it. I thought the water had come from a neighbouring polder; I learnt only later that there had been a major break in the dike at Ouwerkerk and the water was coming from there. And we were floating straight in that direction, with the storm, the northwester, straight toward

> the break in the dike. Yeah, it was in the middle of the afternoon, the afternoon ended, and then it got dark . . . First I saw my brother. Then there was a neighbor. My father and my brother were still floating there, and as it got dark I followed them. Then we could see nothing, nobody could see anybody anymore. I never noticed that I floated through the broken dike, but later that proved to be the case.

He was adrift on a beam in the middle of a storm in the dark, headed towards the North Sea. Day came, 'and then I was in the middle of a huge expanse of water, but it could have been in the middle of a polder . . . or in the North Sea, or in the ocean – water is water. I hadn't a clue as to where I was, because there was no point of reference.'[19]

Regional and religious frames

This description is unique to the person who made it. But there are many similarities to be found in the interviews. Equally striking is the fact that much of the existing written regional literature and many memoirs share stories that are recounted as authentic individual experiences. Apparently it is hard to find words for the story outside the familiar framework borders of the region's folklore. Thus the fantasy of what happened is repeated endlessly. As a result, there are too many eyewitnesses, for example, to the unsuccessful attempt to save the farmer's wife, who was too portly to get through the door of the rescue helicopter. (This particular story was reinforced by the fact that television had recently shown footage of the event.) This tale was confused with that of the woman who was too fat to escape through a small attic window. However, the underlying message of both of these was how hard it was to escape and survive. Even today anyone observant who walks through the villages of that area in Zeeland will note that one need not be overweight at all to have the same difficulty.

Life stories are not just simple narratives. Those who tell them have a difficult time finding the right words to express things that are so personal and so far beyond the reach of the formal language of commemoration. In her book *Narrating Our Pasts*,[20] Elisabeth Tonkin suggests how interviews may be shaped by the expectations of different genres. In my research on the flood, it became clear that among a number of different genres, some secular and some religious, a specific genre was being used – that of the Bible story – or, more accurately, reference to Bible stories. The Bible is never directly quoted, yet even the stories told by non religious people are framed within a biblical context. This language, this way of speaking, was natural for the interviewees, given their strongly religious upbringing and cultural background.

The events of the flood were placed in a religious context from the beginning. That was to be expected, since Zeeland is one of the most religious and strongly Protestant regions of the Netherlands. Practically all of the over 200 interviews conducted made reference to biblical stories, as does all the local writing. Here is an example from the local memorial literature, which uses apocalyptic and biblical language: 'What is that deadening din that comes from afar and rises above the sound of the storm: It is as if the devil rides across Tholen with a thousand coaches, this noise. Frightening is the speed with which it approaches; suddenly it is near. It is as though the earth has split, so great is the tumult.'[21]

One of the books cited by nearly all the survivors, *Toen het schuimend zeenat hevig bruiste*, is a devoutly religious volume from one of the most severely hit islands. Everyone I interviewed from that island was either in agreement with this book or opposed to it. It begins with a sermon given the Sunday before the flood.[22] The local parson tells of a strong intuition that forced him to choose a different biblical passage on which to base his sermon than what the church calendar dictated. The passage he chose was Psalm 119. After the Bible reading he warned the parish of the coming judgement. According to some of the interviewees, the parson said: 'It was not my intention to give this sermon today. The words came into my heart, and I chose a Bible text that reflected them. I do not know what is going to happen; it is hidden from me in the dark.'

Of course sermons about hellfire and damnation were common; in fact they had become a kind of tradition. And in the depth of the winter there were many storms; it was common for water to come over the dikes. In order to identify which parts of the interviews were repetitions of published materials (which in turn are mostly variations of a biblical text), it was necessary to examine the regional literature. For example, the following citation is from the book *In de greep van de Waterwolf* [In the Grip of the Sea Wolf]:[23]

> God saw the work of all men and saw what they did He considered evil. They loved money and power; one people approached another and violence ruled the streets. God loved His creations, and their wickedness saddened Him. Then, He smote with His justice and chastised us, because the whole of humanity should learn that love is greater than violence.

Let us compare this text with Genesis 6.5–7 and 13:

> And God saw the wickedness of man was great in the earth, and that every imagination of the thoughts of his heart was only evil continually.
>
> And it repented the Lord that he had made man on the earth, and it grieved him in his heart.
>
> And the Lord said, I will destroy man whom I have created from the face of the earth; both man, and beast, and the creeping thing, and the fowls of the air; for it repenteth me that I have made them.
>
> . . . And God said unto Noah, The end of all flesh is come before me; for the earth is filled with violence through them; and behold, I will destroy them with the earth.

It is obvious that the story of the biblical Deluge invites comparison with the events in Zeeland. The Bible is the dominant Christian representation of our society. Those who are devout think in biblical terms; they either know parts of the Bible by heart or can paraphrase them. God is there in a very personal way; they can call his name and speak to him.

The story of Mrs M., who told of how she floated on a raft into the next street, has the same tone as Lamentations, which she mentioned several times.[24] She was a member of the most orthodox part of the population, and she was one of the few who talked about a near-death experience. Most who had an experience like hers drowned in the end:

> The raft broke and we were in the water . . . I remember sinking into the water and bumping into wood and all kinds of trash. And I sank . . . I couldn't go back, even though I wanted to. But then I thought – I was raised to be God-fearing – I'm not going to fight any longer. I'll drown and then I'll be with Jesus. Of course I was scared about what was going to happen, but I told myself, just a bit more and it will be over . . . Yet I was absolutely nowhere near drowning. I thought, yeah, this is all frighteningly difficult, but things will work out. Just as if I were a child.[25]

Is this how a child would react to drowning? Or is it more likely the manner in which a religious adult chooses to remember? I think the latter.

It is obvious that the description of drowning and the religious piety that helped Mrs M. were phrased according to her interpretation of the Bible. Its text was inspired by centuries of desperation; one of its pillars is the prayer of the person in need, the person who must look God straight in the eye. It is impossible to interpret these tales of great distress, which grate on one's conscience and attempt to sway the unbeliever, in any other manner. For Mrs M., words other than those from the Bible were probably too empty to describe what she had lived through and survived. Another person I interviewed expressed it well:

> What can be told in an hour? An hour is incredibly short. You can never report everything that happened. That's just impossible. Yes, you lived through it, but to express all you felt and experienced – you can't describe it all. I truly believe that emotions can't be explained. And the pain and the suffering you lived through, you can never . . . well, there are no suitable words.[26]

The Bible (and especially the Old Testament) is one of the most obvious frameworks in which memory is put into words. This is true for both religious and non-religious persons, because the stories are known and shared by everyone within a given culture. It provides a mechanism for understanding which is available to all, and which all seem to interpret in the same manner. In particular, the Bible provides us with words to help give voice to the unspeakable. Often memory is stored using the structure of a Bible story, and literature using this frame of reference is easily accessible for the majority of the population.

Apparently it is difficult to recount such horrible happenings. Human memory becomes fragmented; at times during an interview the residue of other memories tends to surface. In such instances the interviewee realizes the futility of trying to express his or her memories in words. The Bible – with its telling of Exodus, of the unworthiness of humanity in the eyes of God, and its tales of chastisement and testing – gives a symbolic voice to this kind of extreme experience. In my most recent research, I found that beyond the feelings of impotence against the supernatural there are feelings of impotence against nature; and beyond the feelings of being tested (as in the book of Job, for example), there are feelings of being tested by the sea.

During the interviews I conducted, I was never aware of consciously collaborating with those I interviewed by remaining silent during their stories of being tried by God. Neither did I feel that the interviewees were working within a particular genre and giving it a new form, a form in which decades of religious training and thought set the tone. Yet the words were there, and sometimes even the same context. What I must emphasize, looking back, is the nearly frozen and preshaped narrative structure of the storytelling. This means that the genre is fixed, in a way, but each eyewitness has a unique voice that is grounded in personal experience.

The unique tale of each witness turns loose a flood of memories for the teller, and gives these emotional remembrances a healing channel to the outside world. In all modesty I would like to propose that this emotional lancing of memory which takes place between the historian and the eyewitness to tragedy can bring us both a genuine peace of mind.

The perception of threat, of risk, is culturally determined. There was a time in the Netherlands when the danger from the sea was real. In a sense, it is less real now, thanks to the mammoth waterworks, although no authority will deny that another catastrophic flood is still possible. Other threats are perceived as more ominous; nuclear power plants, for instance, are culturally more frightening than the rising sea or certain forms of pollution. Floods are no longer an ideologically realistic threat in public discourse.

People, when interviewed, are aware of that fact, and it all adds to the confusion. Life has changed so much since those days of disaster. They don't know if it was better then or if it is better now. They know how much they have endured to rebuild their houses and their cultural patterns, but they have often lost their sense of community and belonging. The worst moment was when they came back after months and even sometimes a year of evacuation. The house often still stood, and there might even be unbroken teacups in the cupboards. But everything was dirty, smelly and covered in salt. Whole industries sprang up that specialized in the fight against the salt inside and outside the houses. In some cases people started to live on the upper floor of their house before it was dry. Drinking water was restored within weeks. But even months later people were still missing, and sometimes corpses would be found in the drying fields and cleaned ditches. No one talked about that night. Instead they would ask each other how the cleaning and carpentry were coming along.

In the villages a vast process of modernization took place. Indemnity money that came from all over the world created the possibility of electricity and a modern lifestyle, complete with bathrooms and paved roads. The connection with the mainland was improved before the dikes were built. In the upheaval after the flood, survivors experienced the world beyond the dikes. They saw 'peaches in the winter', they ate foods that were entirely new, and they went to schools populated by a different breed of people. (Some of these changes started in a small way before the Disaster, during the War.) Survivors retained pride in their old culture but also took in the new possibilities offered by modern society. Hostility towards the foreign culture of immigrants became impossible for those who had been guests on the mainland. This move to the mainland ended some of the social control of the old society, which had been unbearable for many who had fled to the city. It also put a stop to the naïvety characteristic of such an isolated society. Once the fields dried and were desalted they were marked in new straight lines, altering even the face of the countryside just as its culture was changing.

The isles are a different world, and it is hard to imagine what they looked like before the Disaster. People are silent and scared because of the lack of open speech about the trauma, and, as I have indicated, the confusion is made worse by an overlay of national reminiscence. On days of wind and rain, the same enormous mass of water still pushes against the dikes, keeping fear of the water alive. People still walk to the harbour to discuss the tide, but now they know what they didn't in 1953: that the water can come again.

Notes

1 The material for this chapter has been published in S. Leydesdorff, *Het water en de herinnering, De Zeeuwse watersnoodramp 1953–1993* (Amsterdam, 1993).
2 H. Hartman, 'Het Zeeland onderzoek', enquiry done by the Department of Social Research, Faculty of Social Sciences, Amsterdam, 1991.
3 Kai T. Erikson, *Everything in Its Path: Destruction and Community in the Buffalo Creek Flood* (New York, 1976).
4 Ibid. 178.
5 Ibid.
6 Interview 52.
7 Interview 154.
8 Anonymous interview.
9 L. Lévy Bruhl, *L'Ame primitive* (Paris, 1963), especially chapters VII–IX on the dead.
10 M. Douglas and A. Wildawsky, *Risk and Culture: An Essay on the Selection of Technical and Environmental Dangers* (Berkeley, 1982).
11 J. Berlioz, 'L'Effondrement du Mont Granier. Les textes et légendaires du XIIe au XVIIe siècle', in J. Berlioz (ed.), *Monde Alpin et Rhodanien* (Grenoble, 1987).
12 Interview 214.

13 See R. J. Lifton, *Death in Life: Survivors of Hiroshima* (New York, 1968); also Erikson, *Everything in Its Path*, Chapter VII. For trauma, see S. Leydesdorff and K. Lacey Rogers with G. Dawson, *Trauma and Life Stories: International Perspectives* (London, 1999).
14 Interview 22.
15 Interview 125.
16 Interview 33.
17 Interview 8.
18 Erikson, *Everything in Its Path.*
19 Interview 136.
20 E. Tonkin, *Narrating Our Pasts: The Social Construction of Oral History* (Cambridge, 1992).
21 J. Broersen and T. Koopman, *Stavenisse. Kroniek van een verdronken dorp* (Hoorn, 1953).
22 R. Valkenburg, *Toen het schuimend zeenat hevig bruiste* (Dordrecht, 1983).
23 C. Baardman, *In de greep van de waterwolf* (The Hague, 1953). See also J. Ossewaarde, 'Dan vecht mijn land, het verhaal van de watersnoodramp van 1 februari 1953', unpublished PhD thesis (Amsterdam, 1953).
24 Waters flowed over my head; then I said, I am cut off
I called upon Thy name, O Lord, out of the low dungeon
Thou has heard my voice; hide not Thine ear at my breathing, at my cry
Thou drewest near in the day I called upon Thee; Thou saidst, 'Fear not'.
25 Interview 12.
26 Ibid.

Select bibliography

Baardman, C., *In de greep van de waterwolf* (The Hague, 1953).

Broersen, J. and Koopman, T., *Stavenisse. Kroniek van een verdronken dorp* (Hoorn, 1953).

Douglas, M. and Wildawsky, A., *Risk and Culture: An Essay on the Selection of Technical and Environmental Dangers* (Berkeley, 1982).

Erikson, K. T., *Everything in Its Path: Destruction and Community in the Buffalo Creek Flood* (New York, 1976).

Lévy Bruhl, L., *L'Ame primitive* (Paris, 1963).

Leydesdorff, S., *Het water en de herinnering, De Zeeuwse watersnoodramp 1953–1993* (Amsterdam, 1993).

Leydesdorff, S. and Lacey Rogers, K. with G. Dawson, *Trauma and Life Stories: International Perspectives* (London, 1999).

Lifton, R. J., *Death in Life: Survivors of Hiroshima* (New York, 1968).

Ossewaarde, J., 'Dan vecht mijn land, het verhaal van de watersnoodramp van 1 februari 1953', unpublished PhD thesis (Amsterdam, 1953).

Tonkin, E., *Narrating Our Pasts: The Social Construction of Oral History* (Cambridge, 1992).

Valkenburg, R., *Toen het schuimend zeenat hevig bruiste* (Dordrecht, 1983).

5 'Our land is our only wealth'

Changing relationships with the environment

Olivia Bennett

'In the old days our people knew so much – their knowledge was vast. But it seems as if the drought is drying up the skills and knowledge that we had.' Fatimata Diallo, 60, speaks of the changes she has witnessed in her homeland, Mauritania. Since the 1970s, in the Sahel region of Africa – a ribbon of countries south of the Sahara – lands which were fertile, if only marginally so, have been subject to prolonged drought. The effect on the region has been compounded by man-made environmental degradation. The resourcefulness of the Sahelian peoples, skilled at eking out a living from unfavourable soils, has been tested to the limit. As if this was not enough, people are faced with unprecedented social change, economic and cultural. The result – for many of the 60 million people who inhabit this huge area – is a significantly altered relationship with their environment.

This chapter draws upon a range of individual oral testimonies to illustrate some elements of that changed relationship, and to show how this experience is echoed in other rural communities in Asia (India), South America (Peru) and Africa (Lesotho). The Sahelian interviews were collected from more than 500 mainly elderly narrators, in eight countries and seventeen languages.[1] The other three collections form part of an international oral testimony project focusing on highland communities, inhabitants of complex and fragile environments which are increasingly being affected by development, much of it driven by the demands of lowland urban populations. Again, testimonies were gathered by and from local people in local languages[2].

The interviews were generated by the work of the Panos Institute, a development information organization, which has collaborated on testimony collection with partner organizations in more than twenty countries since 1991. Panos' Oral Testimony Programme takes oral history methodology and adapts it, training and working with small, community-based, often under-resourced groups. Interviewers possess varying levels of education; many have little or no experience of this kind of work. The Programme was developed in response to the fact that those most directly affected by development schemes are often the poor and the marginalized, by definition the less literate, the least influential, the least vocal. Almost always on the fringes of development decision-making, they are nevertheless central to its implementation. As Salme

Dumaneau of Niger put it: 'Our land is our *only* wealth. We are inextricably bound to her. Her poverty is our poverty, we suffer together.'

These experts in the realities of development have had little opportunity to influence theory or practice. The aim of the Oral Testimony Programme is to amplify the voices of those whose knowledge of change and development is first-hand – those living with it, rather than those planning or observing it. Oral testimonies do not replace quantitative or scientific research on the development process, but they do complement and illuminate it. Each collection may be no more than a snapshot of the communities concerned, but the testimonies quoted here represent people's perceptions of environmental change as documented by themselves, not by outsiders. They reveal, to those prepared to listen, what people *believe* to be happening and why. They have provided the communities concerned with moving and compelling evidence to catch the ear of the elite, and humanize the development debate.

Taken together, the range of individual voices in the Panos collections helps to form a vivid picture of rural societies, their changing physical and social environments, and their concerns for the future. Above all, each collection demonstrates the complexity of change in rural areas, and how the social as well as the physical landscape is altering, sometimes at a rate so fast that it threatens any sense of continuity between generations. 'This is a transition period', warns Mohan Lal Uniyal, an Indian village Ayurvedic doctor. 'On the one hand people are losing their traditional knowledge; on the other they are not getting adequate training in modern techniques. As a result they are losing the assets they do have.' For most of the people in question, these 'assets' are the environment; the resources it supplies and sustains; and, crucially, their knowledge of those processes.

This sense of being in transition is echoed time and again in the testimonies. In none of the regions in which Panos has been working has any one factor been responsible. Climate change, deforestation, population pressure, industrial development, modernization, education, migration and changed aspirations – references to all these weave in and out of most people's accounts. Nevertheless, it is possible to highlight key reasons perceived as particularly crucial in the combination of cause and effect which is altering the narrators' relationship with the environment. In the Sahel it would be prolonged drought; in India, deforestation; and in Peru, industry and its dark shadow, pollution. Finally, and more starkly, those interviewed in Lesotho faced forced removal from their land, resettled to make way for a massive reservoir. The reasons for change may be varied, but the consequences have a similarity: displacement, discontinuity and diminishing natural resources.

'A sacred asset'

Most of the narrators felt themselves to have been, as Salme Dumaneau put it, 'inextricably bound' to their surroundings. Many spoke of the land as though it was human. 'Our ground never refuses to accept what we plant.

Even if it is tired it still does its best to produce a small amount', explained Elisabeth Nadjiyo, 45, of Chad. Sometimes land was described as an extension of the family it had nurtured for generations. In the Lesotho highlands, Thabang Makatsela, 57, contemplates removal from his home. What will be hardest for him to leave behind? 'Firstly, my father's big field down the valley which was a source of food for the whole family throughout my childhood. All my brothers and sisters knew we all survived because of that field. This is to say, we grew up feeding on that field.'

But wherever the market economy had made inroads, the value attached to fields was shifting: 'Land used to be considered an almost sacred family asset; today, fields can be sold as if they are just another item of merchandise', laments 65-year-old Kouré of Niger. El Haj Chaï Bou Bagouma, 87, agrees: 'Land . . . was too sacred to be bought or sold. Today, the ground is no longer respected: it has become a saleable product, like any other. Our monetary economy has reduced our land to common merchandise.'

Given this powerful relationship with the land, it is not surprising that many farmers used images of sickness and ill-health to illustrate its reduced productivity: 'The soil has lost fertility; it is no longer allowed to rest. It should be like a patient for a year, then it recovers', commented one Ethiopian village priest. Hadiza Hassane of Niger described the soil at a time when the rains had failed as 'having the pallor of near death'. Farmers everywhere were concerned at the effect of chemical fertilizers, sensing that they were too harsh and forced the land to produce more than its natural capacity, eventually drying out the soil. 'At first I thought it was wonderful, but the fertiliser harmed my land in the same way that a man's body is harmed when he takes alcohol. It was too strong', complained Sudesha Devi, India.

For the majority of older narrators in all four collections, the environment provided them not only with food, but with almost everything else they needed. Mbare Ndiaye of Senegal related how, when he was young:

> there were a great many trees and plants in this area. *Xiile* was used to make the framework for the roofs of our houses, on top of which we put layers of tightly woven *bili* grass. *Jebe* was used for the lintels of certain buildings before the *banco* (mud and manure mixture) was plastered on. Ground *xame* was used for treating sick animals. It was added to their food . . . We used to make most of our tools and clothes . . . But for many years now, people have brought with them ready-made fabric from Europe, so the art of making our own cloth began to die out. There is a great range of goods that we can buy close to home . . . Nowadays, young people don't seem to know the many uses to which trees and plants can be put. Besides, certain plants that we used to gather, such as *saake* which grew in the flood waters, have now disappeared.

While some traditional skills are disappearing, high degrees of self-sufficiency remain where people's limited contact with the wider world allows or dictates it. Asuji Devi, belonging to one of India's Scheduled Tribes, lives in the high valleys of Jaunsar. 'We make [our household goods] ourselves in the village. The large stone utensil for pounding paddy, for instance . . . We make *ringal ki bisai* ourselves (a round vessel for drying grains). We also make bags made up of ropes, different types of brooms, material for plastering stoves and whitewashing houses, [big brushes] and all kinds of things needed to clean the house.' She goes on to list the traditional crops still grown, wild vegetables and fruits, and the natural dyes made from roots and leaves. Yet even she has replaced the latter with synthetic dye for weavings, finding the customers now prefer bright chemical colours to more subtle natural hues.

What people could not grow or make, they would gain through barter and exchange with those who could. The environment was there to be used. People's relationship with it was based on self-interest, for their survival depended on it. Such a profoundly practical relationship does not, of course, preclude a spiritual or aesthetic one – the testimonies contain plenty of evidence of powerful feelings of attachment and reverence. What does become clear, on reading these hundreds of testimonies, is that those in their seventies and older see themselves as the last generation to have such a close and mutually sustaining relationship with their surroundings. Modern life may have brought advantages, but it has loosened the bonds with the land. Jagat Singh Chaudary, an Indian in his forties, recognizes this distinction with regret. 'We have distanced ourselves from nature, while our elders were so closely allied with it.'

The Sahel

Since the 1970s, prolonged drought and man-made environmental degradation in the Sahel have led to accelerating desertification. Pastoralists have seen their herds dwindle and grasslands disappear, and have increasingly turned to farming, encouraged by governments for whom the nomadic lifestyle has always been problematic. One 92-year-old summed up the indignity thus: 'The Peulh have lost most of their animals. They have to drink milk and yoghurt made from powder bought in a tin.' Fishermen, too, have to find other means of livelihood: 'The rain no longer falls and there are no longer any little streams where the fish can lay their eggs before swimming back into the main river . . . Our traditions have been buried in the sand . . . We, the Kotoko people, are a race of fishermen. Now that our fish have gone, what are we supposed to do? . . . We began to rely more on farming' (Abouna Ali of Chad).

As more land is cleared and more intensive farming methods adopted, tree cover diminishes further, exposing the region's fragile soils to even greater heat and wind. Obo Kone of Mali, born in 1912, describes the changes he has witnessed:

> the rain gradually petered out. We began to cut the trees down and lose respect for our old customs. We don't really understand what happened:

> suddenly the rain lost respect for its old cycle . . . Today the environment is sick, the soils are poor and hard, and the trees are dead, having been scorched by the sun . . . I believe these changes can be attributed to the fact that we have lost respect for our customs. We have violated old prohibitions to allow room for modernization and in so doing we have disregarded God's laws.

His concern that some fundamental natural or divine law has been transgressed is one reiterated by many Sahelians: 'Once there were many trees, but now they all have perished. Can anyone other than the good Lord be responsible for this? Perhaps it is because we no longer observe our customs and traditions' (Batomi Dena, Mali). 'Now [the trees] have all gone, and we can only blame God for the crisis. Of course, we do not say that we have not played a part but we are not the main factor. After all, is it man who makes rain? No! Do not point the finger at men. But I do fear for the younger generation who disregard our customs and traditions' (Zouma Coulibaly, Mali).

The references to God's act of withholding rain may sound exactly like the kind of fatalism which exasperated many a development expert and led them to underestimate the understanding and knowledge of peasant farmers. But the value of these oral testimony collections in the development context is that they demonstrate how inadequate a reading that is. The interviews were unhurried, conducted with an interviewer of similar background, at a place and time chosen by the narrators. They show that virtually all speakers go on to demonstrate a detailed understanding of how man-made actions have exacerbated the situation.

Lost respect, lost custom

The majority of the Sahelian speakers were elderly, and some were clearly bemused and disturbed by the by-products of social and economic development, such as the changed attitudes and ambitions of their educated offspring. Yet their sense of regret for lost custom is more than the eternal grumble of those who always think the 'old days were better'. Many of the traditions which have been violated are highly practical methods of environmental protection, which they can no longer afford to observe. They no longer leave the land fallow. Trees which were only valued for particular purposes, for example for medicinal properties, and were never cut down for fuel or building, are no longer protected. Practices designed to maintain the supply of resources have broken down. An elderly Malian woman, Kouahan Sanou, gives one example: 'We sell sap from trees in the market. In the past, because we knew as soon as a tree began to lose its sap it was nearing the end of its life, we did not over-exploit them. Nowadays, there are so few trees, we take sap from any we can find.'

As resources dwindle, tensions rise. Narrators spoke of increasingly strained relations between farmers and pastoralists, who are now competing for resources instead of maintaining the useful symbiosis between their different ways of life. Gawa Assoume of Niger describes how pastoralists contributed to agriculture by keeping their animals in farmers' fields during the dry season. 'This practice benefits each group. The fertility of the land is improved by the animals' excrement, whilst the animals are well nourished by the stems of the crops.' Today, this mutually accommodating relationship is breaking down, as Sadou Iboun of Burkina Faso explains. 'Today pastoralism is an impossible activity because there is no rain and no grass. There are no pure pastoralists left in this area: everyone has land which they farm, in addition to keeping animals. Animals deprived of all good grazing tend to go into the fields and cause damage. This has created a dangerous state of affairs between the farmers and the pastoralists.' Many others bore witness to the rise of conflict, sometimes armed, between these two groups.

'Global warming' or God's will – what is clear is that prolonged drought has combined with other changes to generate environmental degradation at an unprecedented rate. According to M'Bareck Ould H'Meyid: 'We used to say that in Mauritania there were three things to prevent you feeling sad: the water, the grasses and the beautiful views. Today, if we want to lift up our hearts, we have to search hard for these things.' This perhaps depicts a gloomier scene than need be. Many, even the old, recognize that modernization has also been a positive force for change. More problematic, however, is social change, and its most visible manifestation: out-migration. A group of Burkinabé farmers discuss the phenomenon:

> There are the money hunters who go to Ghana or Côte d'Ivoire . . . The other sort are those who leave their villages to go elsewhere in Burkina Faso as they are unable to survive on the meagre harvest from their infertile soil. The density of population has led many of our younger people to migrate. It is only those who have a deep love for the land which they farm who are prepared to stay.

Whether by choice or circumstance, those for whom land cannot be 'the only wealth' have to negotiate new environments.

India

In India, testimonies were gathered in the Western Himalayan regions of Kumaon and Garhwal and Kinnaur, in the states of Uttar Pradesh and Himachal Pradesh. Change is everywhere recorded, as previously isolated communities begin to feel the effects of development. In these mountainous regions, roads are particularly significant precursors of change. They bring access to health care, schools and markets – but also bring crime closer and open up local resources to outsiders, and their construction often threatens the stability of these highland environments. Time and again, in rural areas all over the world, loss of tree cover has accelerated dramatically once roads allow outsiders to penetrate forests previously used only by locals. Bachani Devi, 72, explains: 'Now that the road has come . . . the availability of grass and firewood has dwindled. Now people come from far distant places. They cut grass and firewood and put in on the buses and take it away . . . there is a crisis for us.'

On balance, most welcomed roads, feeling that their isolation and consequent lack of exposure to new ideas and development initiatives had held them back from economic progress. But a minority recognized that unless they were ready to take advantage of the opportunities roads brought, the price paid could be too high. One Himalayan farmer warned: 'People believe that if there are roads in the village, development has taken place. But what is the direct benefit of having a road in our village? The people here are connected not with roads, but with their forests. The grass will go, the trees will go, the stone will go from our village.' He went on to explain that only when they are in a position to market their own goods – vegetables, woven baskets, carved stone handicrafts – would the road benefit them rather than outsiders.

The importance of people's connection with their forests is stated clearly by Lakupati, a tribal woman from Kinnaur: 'Our entire life depends on forests. We get firewood from forests, wood for house construction, and also fodder for our cattle . . . We also get grass, leaves of trees, precious herbs and minerals for our animals. In addition, forests also give us tea leaves, humus, fertiliser, and so on.' Lakupati belongs to one of India's Scheduled Tribes, who retain many of their traditional practices. Elsewhere, links with the environment have become more tenuous. Government forestry policies have sometimes been more of a hindrance than a help, especially where measures to protect trees have been imposed in such a way that local people feel they have lost 'ownership' of their environment. One narrator in his forties described with sadness how many of his own generation have come to feel more detached about the forests:

> Earlier there were dense forests and there were many species. But now in the monoculture pine forests, there is no diversity . . . People had a deep feeling for the forests and when I asked [why] they said they got their vegetables, medicines, as well as grass, fuel, etc. from the forest. If there was [a forest fire] everybody went in a group to put it out. Today it is just the opposite . . . people are indifferent because nothing which belongs to *them* is burning, it is the forest department's . . .

Loss of control over the surrounding resources was a recurrent theme in the testimonies. Many spoke of tensions between government bodies anxious to protect biodiversity and landscapes, and local people's ability to exercise their traditional or customary rights to the resources. Recently, the money-earning potential of certain mountain herbs and plants has been seized upon by the powerful Indian pharmaceutical industry. But as 74-year-old Tegh Singh Mahant found to his chagrin, this recognition does not extend to those who have for years painstakingly acquired knowledge of the plants and their uses. 'I practised in medicinal herbs, so I have some knowledge . . . but we villagers are not allowed to collect [the plants].' The Indian government, correctly recognizing the need to protect this resource, had appointed people to supervise collection, but instead of locals, 'a third person would get the contract to extract herbs. They are really exploiting the treasures of *our* jungle.' Once again, the farmers' direct link with the resources around them has been weakened. One of the more far-reaching impacts of such policies, the testimonies show, is that local people have had to abandon traditional management practices, such as the sustainable harvesting of medicinal plants, leading to an irretrievable loss of indigenous knowledge. Similarly, careful tree lopping practices, once regulated by the entire community, have been given up. Elsewhere, seasonal firewood collection is no longer practised. Why, they reason, should we protect the environment for other people?

In this region of India, deforestation was cited as the primary cause of environmental change. Many spoke of the scarcity of fuelwood; lack of fodder meant numbers of livestock had been severely reduced. And whereas in the Sahel it was the lack of rain which exacerbated deforestation, here people saw deforestation as having brought about a decline in rainfall. 'The forests have been destroyed. From where will the rains come now? Because of the presence of forests, the wind used to move, carrying rain. Now with what will the clouds collide?', asked one old woman. But here, too, there is plenty of evidence that the narrators appreciate that environmental degradation is being driven by a multiplicity of factors: population pressure; increased development bringing roads, housing and more intensive agriculture; and not least, policies which encouraged conversion of mixed forest to monoculture and its commercial exploitation.

One narrator, Jagat Singh Chaudary, believes the imposition of monoculture in the forests has been fundamental to the area's decline. Chaudary's line of resistance – he is single-handedly creating his own mixed forest, seeking to break the stranglehold of the pine – stems from his belief that, without diversity, the forests will die. 'What is being preserved? The pine . . . after twenty to twenty-five years we won't find any soil, only stones.'

'Every house has matchboxes'

Chaudary is in his mid-forties, and is clearly nostalgic for the past. To him, the defining characteristic of the older generations 'was that they were very close to nature . . . There was a sense of cooperation among them. Today each person is isolated and wants to depend on his own resources.' A sense of community spirit, he feels, now has to be coaxed out of people.

> I am filled with amazement at how there is such an enormous difference between now and then . . . the community spirit that [has to be] brought forth today existed in the people naturally earlier . . . For example, [today]

> for canals and *ghul* there is the irrigation department, but earlier . . . if the canal had to be repaired then the entire village went [to do that work]. Earlier . . . whoever lit the fire first, would provide ignition for the whole of the neighbourhood. Today it is just the opposite, every house has matchboxes . . . and although it must have made people self-sufficient, all that talk of community is over.

Although he is convinced that he must resist the decline of his environment, he is doing so as an individual, and in an 'old-fashioned' way, relying on an instinctive feel for what is right for the land. He avoids making grand claims for his work. 'I can't really talk of "experimentation". It's just the truth that I always loved trees and I want more and more species to be planted.'

He contrasts this with the modern approach. 'The generation of today wants returns immediately. It talks of science in books, while the earlier generation practised it on the ground.' For him, the only evidence worth trusting is what you see before you:

> Whatever the books of science say I don't have much faith in that. Because I haven't been planting trees after reading any books, nor have I been trained anywhere or had anybody's guidance. Whatever I have done is on the ground . . . *banj, deodar, bans, surai, angu, chir, bhimal, timla* and *sisam*, etc. will all grow at the same place, if a person is determined . . . [The attitude of] the government officials? Their thinking was bookish, that at this height this species will not grow with that species . . . Yet you have seen them.

Chaudary expresses a feeling which is particularly common among older narrators, untouched by formal education. In all the country collections, they convey frustration that their own environmental knowledge, based on practical experience and accumulated over generations, is undervalued by the possessors of 'bookish' knowledge, that which is written down, tested in laboratories and validated by institutions. At times, this has undermined their confidence in traditional methods. Chaudary, however, is not only championing the value of 'old' knowledge, he is using it to challenge the authority of modern officialdom and the rigid application of scientific methods. 'If someone says it a thousand times that it cannot be done, they can keep saying it – but I have done it . . . In fifteen to twenty years I will make that forest into a mixed forest. But I know that no one from the forest department or the administration will say this.'

A wider view

Chaudary may mourn the loss of the old closeness with nature, but the Indian testimonies reveal that modern threats – in particular large-scale exploitation of natural resources – have engendered another type of environmental awareness. In the Sahel, while the encroachment of desertification is being resisted, ultimately drought is seen as God's business, beyond man's control. In India, however, there is a more pronounced sense that human activity bears the greatest responsibility for the changed environment. And perhaps reflecting this, the communities interviewed revealed a considerable amount of environmental activism. This willingness to fight environmental decline may be intensified by the fact that these highland communities regard much of the change and development to which they are increasingly subject, as benefiting urban, lowland populations. The latter's demands – for water, power, timber and other resources – tend to generate development on a large scale, threatening local people's own resource base, but also ignoring their knowledge base, the richness of which lies in its detail and specificity.

The frustration felt by many of the narrators has been channelled into well-organized resistance to the Tehri dam (the primary beneficiaries of which are to be industry and cities), and to timber extraction, through the Chipko Movement. This grassroots movement was originally concerned with the just allocation of rights to exploit forest resources, but it grew to embrace wider aspects of conservation and environmental improvement.

A significant number of the women narrators spoke of their involvement in these campaigns. For example, Sudesha Devi, 50, participated in Chipko, and in the anti-dam campaign, and for the latter was briefly jailed. She felt that other Chipko members had opened her eyes to the wider implications of deforestation, and showed that what to her seemed just a local problem was of much greater significance. 'We knew nothing before they told us, all we knew was about *our* grass, *our* fuelwood.'

Her comment is significant: everywhere people noticed change in those aspects of the environment with which they were most directly concerned. In all of the collections, women's daily responsibilities meant they were usually the first to pick up on changes in the availability or nature of fodder, water and fuelwood. The impact is soon felt, not least as the burden of collection increases (unless they are able to buy what they can no longer gather locally). Where men's work is directly linked with resources, they too noticed difference. Builders and craftsmen, for example, might notice a decline in particular species of tree. Soil degradation, though, can be harder to perceive – until it begins to manifest itself through reduced crop yields. Those who migrate for work at certain times of the year, or more permanently, might start to put together a bigger picture of change, whereas those left to keep the farms going – mostly women, children and the elderly – had less opportunity to view things from a wider perspective. It wasn't until she came into contact with Chipko members that Sudesha Devi saw how the concern of her own small community with deforestation was shared by countless others, and, if left unchecked, had implications for many more.

So one impact of modernization – a massively accelerated demand for resources – has in fact brought about a wider sense of responsibility for the environment. The subsistence farmer's close but exclusive relationship with the environment is developing into something broader, bringing together a much wider range of individuals and groups.

'A matter of sadness'

Being part of a national movement, however, can cause problems closer to home. Bachani Devi, now a widow of 72, recalled the days she defied her husband: 'My husband was a forest contractor. He cut a huge amount of timber . . . forest after forest . . . He was the major contractor and I was his enemy in this struggle . . . The whole village backed me . . . He never said anything to the [other] agitators. But he was very angry with me . . . We even stopped speaking to each other . . . It was a matter of sadness.' Sadly, despite her sacrifice, she was now seeing her success being eroded as new access roads bring more and more outsiders into the area.

When asked why she had taken up this role, given its particular difficulty for her, she merely said: 'It just came into my mind. I gathered everyone together. I am illiterate. I just got in the mood.' Her downplaying of her initiative and achievement may reflect a tendency, noted by some oral historians, for women to view and present their actions more in relation to others than to take centre stage, and to understate their accomplishments.[3] If so, this was probably reinforced by her feelings of regret and inadequacy about being illiterate. Many of the older narrators, lacking formal education like Bachani Devi, felt uncertain about the value of their experiences and knowledge, aware that traditional practices were being displaced by modern expertise.

Chaudary was a confident man with a mission, prepared to assert the value of past knowledge and practice, but others who had been the recipients of prejudice against their 'backward' ways appear to have been understandably reluctant to divulge much to outsiders, for fear of ridicule. There was evidence in the Sahelian texts that some of the older generation even feel inhibited about passing it on to their 'educated' children. While the Indian narrators had not experienced such extreme change in their surroundings as their counterparts in the Sahel, the older generation express the same sense that social, economic and environmental change are working together in an unstoppable – and for them uncomfortable – alliance. One of the most visible results of this alliance is a significant increase in out-migration, as more and more of the young leave the hills in search of jobs. Ironically, those who have the connections and money to exploit the region's resources are outsiders. As Vimla Devi, 58, one of the women left behind, says: 'Those who get educated go away from here . . . but those who have intelligence are coming up because they have money, and our people are running down there to get money!'

Peru

In the Cerro de Pasco region of the central Andes, the population drift from the highlands to the coastal plains has become a torrent. Here, almost a century of unfettered mining development has devastated the physical landscape, and undermined the economic, social and cultural structure of the *campesino* community who have traditionally farmed the lands. 'Young people leave to find work. They don't go because they hate their land . . . the majority go to escape the poverty . . . if there were better pastures and clean water they'd stay', said Delma Jesus Flores, a mother whose own son has left for Lima. Opportunities for employment in the mines have shrunk as a result of privatization and mechanization, yet the polluted pastures can no longer sustain the once substantial herds of livestock.

Industrial development in an agricultural area is always going to bring significant change. The mines brought new work opportunities, the cash economy and improved infrastructure – schools, roads and electricity supplies. They attracted labourers from outside the region, and this in-migration brought new cultures, ideas and customs which enriched local life. Many narrators, not just those who gained jobs, acknowledged these positive changes.

But when industry is developed without regard for the long-term impact on the environment, more drastic change occurs. Inadequate environmental controls, and failure to enforce the few which existed, meant mine waste seeped unchecked into the streams and rivers, and toxic fumes escaped from the smelter, damaging the health of communities, and the livestock on which most farmers depended.

Recognition of what was happening was hindered by the fact that many *campesinos* took jobs in the mines, and this change in occupation in itself altered, and indeed weakened, their connection with their own environment, as Juan Santiago explained. Farmers work according to the seasons and the weather. As a miner, however,

> You . . . are always in the grip of time. You have to start work at a certain time and finish at a certain time . . . they squeeze the juice out of you, you sweat like crazy and come out exhausted . . . It means all you do is work, be at home, perhaps one day you'll go out for a bit and that's it – you're practically enslaved . . . But . . . when I was a *campesino* . . . Although things were hard, I didn't have any money, at least I was free to move around.

Don Hilario Meza Alejandro agrees: 'Yes, it was a change [working in the mines] . . . life in a [*campesino*] community is always different, it's more peaceful, calmer . . . The hours are more rigid in the mine, if it rains, thunders . . . you still have to do the hours and, if you don't do them, then you're punished.'

He went on to describe how this brought about a more profound change to the workers' mindset, a short-termism and individualism at the expense of the wider social and physical surroundings:

> [Miners] think of the present and that's all, as they say. They think about their money, about their work and that's all. They're trapped by their

> surroundings so they're not even interested in whether their lungs are being infected . . . or if they're going to die in the mines . . . That's what happens with contamination, they don't consider that it's going to affect the community later . . . the truth is that when I was young I said, 'I work and earn my money and that's it' and I wasn't really aware . . . I felt more like a miner than a *comunero* . . . But then you begin to realise the seriousness of the problem and . . . you finally realise that you're a *comunero*, because this is your land and you're going to die here. So you do something about it, because they are contaminating the land and, as you can see, sometimes the damage they do to the land is irreparable.

As livestock herds have been decimated, it has become harder to maintain the communal working practices which were such a strong feature of life in the region. Some other social and cultural practices have changed or disappeared, too. Washday was more than a domestic chore, it served an important community function, as Luis Celis, in his mid-60s, recalls with sadness:

> There are things that are no longer customary around here, for example . . . the *faenas* (communal work) of the women washing their clothes on Sundays, a delightful custom that we had before, señor . . . On Sundays families would go down to the river with all their clothes to wash. The whole family and all their friends. But that's when the waters were clean. They would spend the whole day there doing their washing, everyone would help each other out, everybody would be partying. They'd bring a lamb along and they'd cook the lamb on a spit . . . they'd celebrate with their maize wine, their *chicha*, the children would be playing . . . It was really beautiful, señor, the whole day was beautiful . . . [But it doesn't happen] nowadays, no, because no one wants to wash their clothes in the river – the river is pure filth.

While industrial pollution is the main cause of environmental change in this region, as elsewhere, other factors have played a part in the land's declining fortunes. Again, the devastation of their environment has forced the inhabitants into thinking about it in a wider sense. One of the local interviewers, Jaime, commented:

> The older generation were more isolated and tended only to know these mountains. This isolation can breed a kind of acceptance of fate – 'it's natural for the fumes to be here'. They love the land and resist moving away from it more than anything. The young people have not had the same employment opportunities, and they see progress and development everywhere but where they are. They are sold an image, an ideal. Everything is conditioned so that they want to get out – and they are not given a true version of what is happening to their environment.

The interviewing team mounted an exhibition using extracts from the testimonies to promote public discussion on the issue and asked the public to contribute their own memories of the land, and ideas for improvements. People of all ages responded with writings, photographs and paintings. By pooling their different personal perspectives, the collective narrative of environmental change was strengthened, and a network was created bringing together the different interest groups: miners, union members, *campesino* communities, schools, students, mothers' clubs and the retired. They continue to fight for better controls, more compensation and for measures to restore the least damaged land.

Lesotho

The final collection of testimonies documents a more abruptly changed relationship with the environment – forced relocation. The highlands of Lesotho, one of the world's poorest countries, are the site of one of Africa's most costly and complex engineering schemes, which will divert water from Lesotho's rivers to feed the industrial heartland of South Africa. In 1998, villagers from a small number of communities due to disappear under the waters of the Mohale reservoir were moved from their homes and resettled in a number of different locations.

As resettlement schemes go, this was small-scale. Some of the less visible and long-term changes – social as well as economic – which these communities will undergo will not be apparent for some time. Panos and its local partner organizations plan to collect testimonies at different stages in the people's adjustment, to increase understanding and awareness of the impacts of resettlement. The first testimonies were gathered when removal was imminent, and naturally peoples' feelings of anxiety and uncertainty were running high. The testimonies describe the fraught process of resettlement, and its impact on the personal and community sense of self and cultural location.

Perhaps one of the more striking aspects has already been remarked upon – the way that people felt they were leaving behind part of their family: the fields which had nurtured them for generations. And many were more literally abandoning family members, who had been buried in the hills. Although provision had been made to relocate as many of the graves as possible, people weren't sure where some of the older burial sites were and feared their ancestors' distress and condemnation. Tlali Mokhatla, a man in his nineties, said: '[What I shall miss] are the graves which I see that we are going to be separated from and leave [behind]. Even the [dead] are going to rise against us and say "You leave us here, so we could be smothered by water?"'

Thus the environment was steeped in personal associations and history. Here is one old woman's lament:

> Here where I have built, is a place where I have lived well . . . I was ploughing, I was eating and getting full in the stomach. I was planting

> each and every single crop in the fields. I was getting wild vegetables that have been created by God on the ground and I was being full in my stomach . . . I was living comfortably in this land . . . It will remain as a rock on my heart when I think of the place that I am being removed from.

Yet the truth is this area had not been settled for more than several generations, and it had always been a struggle for the people to live entirely off the land. Many of the men had laboured for long periods in the South African mines. Some of the nostalgia evident in these accounts probably can be attributed to people's understandably heightened emotions and trepidation prior to removal, but it may also reflect their concern about a more fundamental and frightening loss: that of independence. They prided themselves on their ability to survive in a relatively harsh environment. As long as a family had animals, they could eat and drink, and plough other fields in exchange for what they could not produce. As long as a family had fields, they would never be totally dependent on others. The men who had worked as miners seemed to regard it as hardly worth mentioning that of course they came back to the land. Useful as a wage was, it offered no security. Someone else made the decisions to hire and fire. Only land represented that crucial degree of self-reliance, and conferred some sense of control over your destiny: 'Ache, the life of town is heavy . . . if you do not have money, you have no means of eating . . . Here at home, the means are many. I use soil, I sow. There will germinate vegetables, maize, potatoes, pumpkin. I eat and become full. I do not buy food', said Mokete Mohaieane (64).

With cultivable land so scarce in Lesotho, many of the resettled had little choice but to take cash compensation, rather than the 'soil-for-soil' they wanted. No one described cash the way they described land. All saw it as a finite resource, something you spent, that would soon disappear, that could even be stolen. Most were unfamiliar with large sums of money, banks or savings accounts, having led a hand-to-mouth existence ('here you catch a grasshopper straight to the mouth'). It would take more exposure to the monetary economy for them to see cash as a potentially productive resource, capable of generating a secure future, the way they viewed land.

'The wisdom of living in that place'

As in the other accounts, it wasn't just the land itself which was valuable, but their familiarity with it. In complex and diverse mountain environments, where steep slopes mean one field can present a variety of climatic and soil conditions, knowledge of the environment is especially key to survival. Sebili Tau summed it up: the greatest loss of all to the villagers will be 'the wisdom of living in that place'. He went on to enumerate the wisdom he had accrued over time – some passed down, some found for himself: the most favourable time and places to find or grow food, whether wild or cultivated, and where to locate medicinal plants, shelter livestock, find the first spring pastures, and the raw materials for building, carving or weaving. If you were lucky enough to be resettled to new fields, he went on, 'once again you [will have to] struggle to learn that land, not knowing . . . at what time [to] look for [the plants] and when they germinate, and so on'.

Thus it was a family's knowledge and skilful use of the land which primarily defined their wellbeing. Similarly, the support they gave or gained from others in the community was closely tied to use of the environment. Cash was not absent from these villagers' dealings, but repayment for help at a crucial moment might well be in the form of some particular seeds at germination time, or when livestock had produced offspring, or by shouldering a skilled or labour-intensive task for someone. Sebili Tau again: 'truly I usually plough in partnerships, right here with the old people or here with people who do not have cattle, or with those who are needy in the hands, like the handicapped'. Thus repayment for favours was complex, and often tied to agricultural rhythms, but the system did rely on a degree of certainty about the world within which it operated. As relocation approached, these reciprocal relationships began to falter and break up. Those who had placed their future in cash sums started selling livestock. Fields were abandoned if the harvest was likely to be post-relocation. And so displacement, long before the event itself, starts to pull apart the informal social support systems which are essential in the daily economic life of the poor. The elderly, such as Nkhono 'Maseipati Moqhali, were particularly vulnerable: 'This place where I am going, what am I going to eat? Who will give me a field? . . . We are separating from our friends, these ones who were looking after us [saying] "Grandmother, take some porridge" . . . It is cruelty.'

'Nature is not as generous as it used to be'

The testimonies from each of these projects leave no room for doubt over the extent of environmental change. For most narrators, the land was their prime asset, their 'only wealth'. Their relationship with the environment permeated every aspect of their lives, and reflected their need to make it work for them, to nourish it in order that it could continue to nourish them. Mostly, but not exclusively, people narrate a tale of diminishing natural resources. As Wakary Gassama, a Senegalese in his seventies, said, 'Nature is not as generous as it used to be.' Some changes have been gradual, others more dramatic. Some are a result of deliberate policies, often formulated by people far from the sites where their ideas take effect; others are the result of more local action.

The testimonies highlight a range of emotions and reactions. People express concern and anger, where external interference has been primarily responsible; bewildered acceptance, where the fundamental cause seems more mysterious (climate change); and rueful responsibility, where they have no choice but to adopt what they know to be short-term policies. But a common thread is that something has been lost. To dismiss the many references to a past 'golden age' as merely uncritical, unreflective nostalgia would be to miss the rich lessons of the testimonies and to underestimate those who voice them. For they are expressing something difficult, conceptualizing a feeling that modernity – with its many, measurable and acknowledged advantages – comes at a high price. State intervention has undoubtedly affected traditional resource management and monitoring – and one result, which emerges from these narratives, is that people feel they no longer control the resources on which they still depend, albeit no longer exclusively. The significance of this lost connection with the environment is as hard to measure as it is to articulate, but clearly it is a source of unease.

Everywhere migration has been one response to changed circumstances: sometimes undertaken in the spirit of ambition – the search for a better life – but often with deep regret. There are examples of adaptation with significant cultural ramifications, such as pastoralists becoming farmers in the Sahel, livestock herders becoming miners in Peru, and highland farmers becoming urban dwellers in Lesotho. And there are examples of people who have stayed with the land and taken what actions they can to reverse the decline or to conserve what remains, and at the same time to preserve their present way of life. And finally, there are stories of individual and collective resistance to the wider forces altering the social and physical landscape. And particularly as a result of the latter activities, some narrators have moved beyond the local to an awareness of the broader implications of environmental change.

For some, the fight to gain greater acknowledgement of the validity of local knowledge – 'the wisdom of living in that place' – is an important part of their fight to halt environmental degradation. For the tendency to mourn the past is, in large part, a reflection of people's perception that their own environmental skills and understanding are no longer regarded as relevant. Many of the narrators have lived through a time when their knowledge, based on hard reality, accumulated incrementally, rarely written down, but crucial to their survival and the wellbeing of their communities, has been superseded by a different kind of knowledge. This is more scientific, based on logic and experiment, research and discussion, publication and peer review, and its practitioners have held the balance of power. A major lesson from these testimonies is that expert research, and the development processes it informs, can only be enriched by taking greater account of those whose knowledge has accrued from first-hand experience of ordinary, awkward everyday life.

Notes

This chapter draws on interviews gathered in many different sites. Thanks are due to all those who worked tirelessly to record and transcribe testimonies, in often remote areas, and with limited resources. Above all, Panos owes a great debt to the narrators, who generously gave their time and their stories, not least in order that their voices might reach a wider audience and contribute to a more inclusive exchange of views.

1 The Sahel Oral History Project was conceived and run by the international development organization SOS Sahel and its partner agencies. An edited selection of the testimonies was published by the Panos Institute: N. Cross and R. Barker (eds), *At the Desert's Edge: Oral Histories from the Sahel* (London, 1991).

2 The three testimony projects with highland communities referred to in this chapter were undertaken as follows: in Peru, Panos worked with a number of communities through the Instituto para el Desarollo de la Pesca y la Mineria, and CooperAccion; in India, testimonies were collected through the Himalaya Trust and coordinated by Indira Ramesh; and in Lesotho, testimonies were gathered primarily through the Highlands Church Action Group and were coordinated by Dr Motlatsi Thabane of the Department of History, National University of Lesotho. The testimonies will be published and broadcast in local languages, as well as disseminated regionally and internationally. The quotations in this chapter are taken from English translations of the original transcripts.

3 See, for example, G. Etter-Lewis, 'Black Women's Life Histories: Reclaiming Self in Narrative Texts', in S. Berger Gluck and D. Patai (eds), *Women's Words: The Feminist Practice of Oral History* (New York, 1991), 43–58.

Select bibliography

Coleman, P., 'The Past in the Present: a Study of Elderly People's Attitudes to Reminiscence', *Oral History*, 14(1) (1986), 50–9.

Cross, N., *The Sahel: The People's Right to Development*, Minority Rights Group Report (London, 1990).

Cross, N. and Barker, R. (eds), *At the Desert's Edge: Oral Histories from the Sahel* (London, 1991).

Etter-Lewis, G., 'Black Women's Life Histories: Reclaiming Self in Narrative Texts', in S. Berger Gluck and D. Patai (eds), *Women's Words: The Feminist Practice of Oral History* (New York, 1991), 43–58.

Gaventa, J., 'The Powerful, the Powerless, and the Experts', in P. Park, M. Brydon-Miller, B. Hall and T. Jackson (eds), *Voices of Change: Participatory Research in the United States and Canada* (Connecticut and London, 1993), 21–40.

Thompson, P., 'Life Histories and the Analysis of Social Change', in D. Berteaux (ed.), *Biography and Society: The Life History Approach in the Social Sciences* (London, 1987), 289–305.

6 Using community memory against the onslaught of development

A case study of successful resettlement in Zapata, Texas

Jaclyn Jeffrey

> The law condemns the man or woman who steals the goose from off the common,
> But lets the greater villain loose who steals the common from the goose.
>
> Old English folk saying[1]

One of the sad facts that development anthropologists, economic and environmental analysts, and other scholars have learnt about large-scale development projects of the past fifty years is that the human costs of such projects often exceed any economic benefits derived from them. In a typical development scenario, such as construction of a hydroelectric dam, communities – the people, their homes, churches, businesses, schools and public institutions – are uprooted and resettled to a less desirable environment. A displaced community is stripped of that delicate interweaving of social, material, environmental and spatial relationships that combine to create viable settlements. Resettlement almost always results in separating the affected population from their land and from access to other traditional resources. Ultimately, and in nearly every case, it leads to impoverishment and social disintegration.[2] This, however, is the story of one community that experienced development-induced displacement and managed to beat the odds, a community that not only survived displacement but lives to tell about it.

In 1953 the US–Mexico border town of Zapata, Texas was relocated to make way for the Falcon Dam and Reservoir being constructed on the Río Grande/Bravo. The US government policy for resettlement was inadequate and deplorable, but the community held together in spite of these divisive forces. Through the telling and retelling of their community history as they go about their daily lives, the townspeople of Zapata have maintained social integration in the face of forces in the dominant society that oppose it. By vehemently preserving an accurate account of their displacement experience, they not only condemn the villain who stole their common but also use the memory of that villain as the underlying focal point for maintaining social cohesion.

The community memory

The displacement story that Zapatenses tell usually contains the following elements. First of all, it does not begin in 1953. The story *always* begins in 1750, when Don José de Escandón, under orders of the viceroy of New Spain, established the colony of Nuevo Santander along the banks of what we now call the Río Grande or, in Mexico, Río Bravo del Norte. The colonists established ranches on land grants issued to them by the king of Spain. At this point, the storyteller will usually add, 'And my family has lived here ever since.' Thus the setting for the story and the rights of entitlement are established. The story then takes a two-century lurch to the 1950s. At that time Zapata was a small town of around 1,500 people, nestled on the banks of the river but essentially isolated from the mainstream populations of both the US and Mexico. It was the county seat, however, and so around the *placita* (small plaza) were the county courthouse and jail, one bank, a hardware store, one school, two churches, a funeral parlour and a pool hall. Many of the buildings were colonial vernacular architecture similar to that found in central Mexico – constructed of quarried native sandstone 50 to 60 centimetres thick, with high ceilings, tall windows, massive wooden doors and *chipichil* roofs, which kept the rooms cool in the hot, dry climate of the *frontera*.

In the 1950s the people who lived in Zapata were mostly farmers or ranchers. The farmers grew tomatoes, onions, peppers, melons, corn and some cotton in the irrigable land along the river. Ranchers grazed cattle and horses on ancestral rangeland, nourishing their livestock with river water, riparian vegetation and thorn scrub. Some of the women did piecework by correspondence for national stores like Sears and J. C. Penney's. The women in Zapata were famous for their fine needlework. At this point, a smocked christening gown or an embroidered quilt stitched by a grandmother or great aunt might be carefully extracted from an old wooden chest to show the listener. No native Zapatense performed migrant labour. Every child attended public school and instruction was primarily in Spanish, which was the everyday language.

Then, the narrative continues, in 1952 men from the federal government came to town and told them that they would have to move to make way for a dam being built on the river. But they were not to worry, the men said, because the government would pay them for their lost property and the dam would provide flood control and hydroelectricity and the lake would be a big boon to the local economy by fostering tourism. However, the people were not impressed with this argument, so the government was compelled to promise to build a new town with modern schools, parks and civic buildings. Before anyone could argue further, government workers began relocating the cemeteries. Ancestors and loved ones were unearthed and reburied in a new cemetery on a hill overlooking the old town site. There would be no more debate about the move.

Appraisers from outside the region came in to assess the value of each structure in the old town and to pay each owner 'fair' market value for their property. It quickly became clear that fair market value was not anything approaching replacement value. And what did these outsiders know of property value on the border anyway? When assessing land value, they ignored the significance of proximity to the river because they failed to appreciate the importance of water in that semi-arid climate. They mistook plastered stonework for adobe, and century-old elegant homes of Spanish colonial architecture were assessed at values lower than small frame houses because they were 'old'.

> Can you imagine anybody telling you that the life span of a home is only forty years! That's what the regulations had. They had homes there that were a hundred years old. They were made out of rock. They [the government appraisers] would just sit there and say 'Well, heck, your life span of your home is only forty years. You've already lived in it thirty-five. You've only got five years left.' They took advantage of the town. My house was stone, twenty-four-inch walls. They dynamited it. The stairways are still there. It was a two-story. I still go over there. (Gabriel Cardoso[3])

Stripped of their ancestral homes, Zapatenses seemed destined to be stripped, too, of the means of regaining their community identities.

Before property negotiations with everyone were completed, the dam was finished, the gates closed, and the lake began filling up sooner than expected. Many people were flooded out of their homes and forced to live in Red Cross tents for as long as eighteen months. Despite government promises, the public water supply, the sewage system, the new courthouse and the new schools had not been constructed – although they eventually were. The government had provided no compensation for moving costs. Those Zapatenses who were able to move their houses were not prepared for the many hidden costs involved – preparing the foundations, realigning the house, re-establishing plumbing, electrical wiring, etc. The stone houses could not be moved so were replaced with modern tract houses. A number of Zapatenses moved off to live with relatives in other towns because they could not afford a house in the new town. Some farmers who had worked all their lives on their own land were forced to become day labourers on neighbouring ranches and distant farms. Worse still, eventually the people realized that the dam had actually been built to provide flood control and irrigation not for Zapata but for powerful Anglos[4] downriver who would transform the arid Lower Río Grande Valley into rich cropland. With great bitterness, the storytellers point in the direction of the Valley: 'People say there was a great "miracle" that turned the Valley into an agricultural paradise. There was no miracle. We were simply sacrificed so they could get rich' (Gregorio de la Peña). The second-generation tend to be more blunt:

> You know who got all the benefit? Lloyd Bentsen [former US Senator and Secretary of Treasury], the son of a bitch! Excuse the words. For his

> family. They had Camelot built eighteen miles away from the river with all the water rights, when they took all our water rights away from here. (Armando Díaz)

Thus is the story of the displacement of Zapata that is perpetuated in the community today. It is almost always factually accurate and has not yet acquired the highly crystallized style which would suggest its entry into the realm of oral tradition, no doubt because many survivors are still living. Placing their story into a broader context, we can make further observations. For 200 years the inhabitants of Zapata did a reasonably good job of taking care of themselves, of making a life along the banks of the river without dependence on any agency. In fact, because of their low water and energy needs, they did a better job of managing resources than most populations in the US. Nestled into the semi-arid biotic region known as the Tamaulipan thorn scrub, they created an economy and style of living that harmonized with the hot, dry climate and harsh terrain. Situated on the periphery of the economic, social and political spheres of both the US and Mexico, the region developed not only a distinct economy but a distinctly rich culture as well.

Life along the *frontera* has always been hard.[5] The countryside is dry and rugged and has for most of its history been isolated from the societies of both Mexico and the US. Many a pioneer came there and left, driven off by the Indians and by the unyielding environment. Throughout its history, from Spanish colonial settlement to the present, the community of Zapata has depended on strong family ties and opposition to outside forces to maintain social cohesion. They have survived onslaughts of Apaches, Comanches, the Mexican army, Texans, the US army, European immigrants and Anglo entrepreneurs. Descendants of the original land grant families have intermarried with Anglos but, remarkably, maintained their own culture. Among themselves, there are factions and bitter feuds, but to the outsider they present a united front. In this way, they represent well the 'persistent peoples' described by Edward Spicer[6] when he observed that groups of people define themselves in opposition to other groups and that identity is forged out of struggle and resistance.

US resettlement policy in 1953 was market driven and inadequate to restore Zapatenses to anything approximating their former individual and community lifestyles. Infrastructure in the new town site was minimal and overlong in being provided. Moving costs, increased cost of living in the new town and loss of valuable farmland depleted any monetary reserves that most of the people had accumulated, making it impossible for them to invest in any new enterprise. Farmers and ranchers who had nourished their families, their crops and their livestock with free river water for 200 years were now forced to apply for permission to purchase that water. They lost their riparian rights when they lost land adjacent to the river, so they organized an effort to compensate for that loss by creating an irrigation district. In what seems to have been a particularly cruel twist, the government refused to grant permission for the creation of an irrigation district. Because there is little subsurface water in Zapata County and little rain, the result was that farmers and ranchers lost access to their primary, and often *sole*, source of water. Those who could afford to do so leased land along the perimeter of the lake in order to water their livestock. Thus, many ranchers found themselves in the nearly intolerable position of having to lease back from the government land they had been forced to sell in the first place. Moreover, the dammed river slowed the flow of water, which increased water contamination. Medical personnel began reporting numerous cases of illness among babies and small children and the source was traced to the water supply. Although the reservoir was held up as a potential source of income, only one relocatee was able to purchase land along the shoreline of the lake – the only land of known potential value after the rich bottom lands of the river were inundated – nor do they own any lakeside property today. Enterprising appraisers and realtors from outside the county purchased and developed lakefront property.

Since 1953, Zapata has experienced an influx of outsiders – 'Winter Texans' (retired persons from northern states who come to winter in the warm climate) and others, Anglo and Mexican-American, who come from 'outside' to operate motels, restaurants and fast-food places which cater to tourists who come to the lake to fish. With a few exceptions, these newcomers have created an opposing force within the community. Mostly but not entirely Anglo, they tend to hold to the 'change is progress' type attitudes and lifestyles of the dominant culture in the US and to denigrate the Spanish-speaking, less competitive, place-attached and family-orientated natives. Native Zapatenses harbour bitter memories of their resettlement experience and are resentful and suspicious of outsiders who come in to take advantage of the lake which was the source of their sorrow. An increasing bitterness is the fact that Zapatenses are being discouraged from preserving these memories and passing them to the next generation. Newcomers want to manipulate the history of the town to conform to one which justifies their own world view in general and their right to live in that place in particular.

Distorting the memory

The experience of Zapata is typical of many displaced populations around the world. A common occurrence in large-scale development projects is the giving away of ancestral lands to outsiders and the incorporation of previously isolated communities into the national economy. This process separates the affected population from their traditional means of production, usually forcing movement from subsistence and petty commodity production to wage labour and/or cash crop production. In other words, development-induced displacement accelerates the transition to capitalism process. Since Marx, and in a long chain of distinguished scholarship, the study of a community's transition to capitalism has tended to be simplified into a 'before-and-after' proposition, from the natural economy of the peasant to wage labour in a market economy. William Roseberry has pointed out that the process is not so clear-cut and suggests the need to look at a community which has severed the connection with a 'peasant past' but attempted to maintain a connection with its 'indigenous past' and present.[7] In the case of Zapata, resettlement linked the local economy with the national and regional, and thus accelerated transition to more intense forms of capitalism than had existed in the community before. Many Zapatenses lost access to their traditional means of production and were forced to move from subsistence and petty commodity production to wage labour. However, Zapatenses are not only attempting to maintain connections with their indigenous past; they are, so far at least, succeeding, even as their local economy has been assimilated into the national market economy. They are able to succeed, in large part, because they control the history of their community.

It is commonly agreed upon by scholars of memory that individuals shape their recollections of the past to meet the needs of the present. Political economy also moulds community memory to its own purposes. For example, Ana María Alonso has shown how the Mexican state 'has "eaten" and "regurgitated"' the image of Pancho Villa and his role in the Mexican Revolution to create an ideology that legitimates the current national government.[8] In another case study in Mexico, Néstor García Canclini examined the way that capitalism has altered folk art in Mexico.[9] Since a community's commonly agreed upon history is a form of folk art – that is, the creative expression of ordinary people in their daily lives – it is appropriate to apply García's methodology to a study of the impact of United States economic development on community history.

In his study, García laid out in detail precisely how folk culture is moulded in capitalist societies. A market economy categorizes, homogenizes and distorts folk art to make it palatable to the dominant culture and convenient to the political economy. According to García, during the assimilation process into a market economy, folk culture is reconceptualized to make it receptive to capitalism. Simplification processes, which suppress plurality and abolish differences by reducing features peculiar to each community to a common type, perform this task. This assimilation does not always destroy traditional cultures; sometimes it merely appropriates and restructures them. Traditional history is rewritten to complement the history of capitalist culture. The meaning and function of traditional objects, beliefs and practices are reorganized, which usually results in distorting and/or destroying the structure of meaning and the sense of identity in the traditional culture.[10]

Over the course of my three years of research there, I noted several examples of how mainstream American (i.e., more capitalist-orientated) outsiders have rewritten the history of Zapata. One Anglo woman told me that she was tired of hearing the old Zapatenses talk about the resettlement experience because they were 'not even citizens at the time anyway' (Ellen Lewis). The implication seemed to be that they are not entitled to begrudge their treatment by the government because they were not legally entitled to live there in the first place. In fact, not only were they US citizens, it is probable that the Zapata families to which she was referring had occupied their homes for a hundred years before this woman's ancestors ever left Europe. As an outsider, however, this woman maintains a viewpoint that allows her to feel as entitled to live in the town as any native. She and others like her – and in a long tradition of colonial thinking – have recast Zapata's history to conform to their need for a world in which their group is dominant by entitlement and for a world in which there are no innocent victims.

In a similar vein, newcomers often suggest that the community is better for their being there. For example, they have told me that if it weren't for the Anglos, there would be no public library, modern medical services, decent schools or good roads. In fact, the public library was established by local ranchers and members of the Soil and Water Conservation Board. The library board of trustees even printed and still distributes a booklet of newspaper clippings about the displacement entitled 'Refugees in Their Own Land.' As for schools, roads and medical services, comparing the county with others in the state suggests that these improvements would have come about anyway as the state and county upgraded facilities and services over the years.

Numerous outsiders refuse to acknowledge that Zapata was a Spanish colonial town and actually have told me that it was little more than 'a few stick houses around a plaza' before the dam. The implication here is that there *was* no community worth saving. Unfortunately, this distortion of town history runs rampant largely because the physical evidence necessary to dispute it no longer exists. There are few photographs of the old town. Just prior to inundation, the government dynamited the stone buildings of Old Zapata, and over the years many citizens have taken the crumbled remains for keepsakes. When the lake waters recede, as they tend to do about every ten years, only the old *placita* and bridge can be identified.

Perhaps the most virulent of historical distortions is the one which claims that Zapatenses are just a bunch of 'cry-babies', that displacement for the sake of progress happens everywhere and moving is no big deal. This theory is especially popular with outsiders who now live in Zapata, probably because the very fact that they are there implies an inclination to favour mobility. These outsiders are movers by definition, and their interpretation of resettlement is well founded by the dominant culture. One reason that attachment to place has been underestimated in development projects is because it is not a concept familiar to the United States. Ours is a nation built by immigrants, with an economy dependent on the mobility of labour. Our whole history is based on the idea of resettlement being for the greater good. We also have inherited powerful Old World metaphors – the Exodus, the Holy Grail, the hero quest – which shape our attitudes towards moving. Moving is opportunity, and specifically *economic* opportunity, which tends to outweigh consideration for home, family or social relationships. The dominant culture of the United States, in the public sphere at least, is psychologically immune to the problems of resettlement.

Maintaining social cohesion

Besides distortion, capitalism affects traditional culture when the economic base of the people is separated from its cultural representations. The unity between production, distribution and consumption is broken, and then these scattered segments of the economy are reconfigured in a way that corresponds to the goals of a capitalist system. This process, of course, often results in severing the bond between individuals and their communities. While many farmers and ranchers in Zapata County lost access to the means of production – irrigable land and rights to water – an unrelated local phenomenon in the 1960s mitigated the loss. Large deposits of oil and natural gas, first discovered in the 1930s, began to be exploited to the extent that Zapata County became the number-two natural gas producer in the state. Farmers displaced by resettlement were soon able to obtain well-paid jobs in the oil and gas fields without leaving the county. Income from natural gas production enabled local ranchers to keep their land, even though many lost access to free water and ranching was no longer the viable business it had been. Oil and gas companies became the largest employers and taxpayers in the county, providing the tax base for good schools and county services, spurring growth of local service businesses, and providing jobs for young Zapatenses so they did not have to leave home.

The town economy blossomed under the auspices of an industry which brought in few outsiders but provided considerable wealth to many natives. The upshot has been that Zapatenses have managed to preserve their community history in the face of distortions by the dominant culture because they had the wherewithal to maintain their economic and political independence. This material factor, coupled with the strong culture of the *fronterizo*, has allowed Zapatenses to manage their assimilation into the dominant economy and culture, to hold on to their indigenous past even as they move beyond their peasant past.

How they have succeeded in preserving community memory can be seen by examining four institutions in the town: the county government, historical commission, local newspaper and chamber of commerce. Local government is on the county level and is controlled by native Zapatenses, who have maintained political control through the traditionally strong *patrón* system. Local officials and the voting majority have consistently rejected state or federal government interference in economic development. In 1964, they flatly refused a massive federally funded project to develop the lake and town area for tourism. One Zapatense businessman, a respected leader in the community, had this to say about the outsiders who wanted to develop the lake for recreation and tourism: 'I used to hate a lot of those guys when they came in here. They always began, "You know what's wrong with Zapata?" They thought we were inferior because we didn't get into the fishing and tourist business' (Roberto Mendoza). Local government provides no enticements and minimal conveniences for tourists. Public recreational facilities are orientated away from the lake. There is no public park area along the shore and only one public boat ramp in the whole county, and it is rendered useless when the lake level drops more than about 10 feet, which happens frequently. In 1995 the county commissioners refused to extend the boat ramp, even though the chamber of commerce pleaded with them to do so and offered to pay for it themselves.

The Zapata County Historical Commission, with the accompanying Zapata County Historical Society, is second only to the Catholic church as the official bearer of local culture. Historical commission members are appointed by the county judge, and they too – with rare exceptions – are natives, sons and daughters of the displaced who are dedicated to the preservation of community memory. Being a member of the commission carries considerable prestige. As one county official told me, 'One of our greatest resources is the history of Zapata' (Juan Chapa). The commission has some mandated functions and restrictions established by state law, but the range of its activities far exceeds those of other county historical commissions. Members of the society maintain the local history museum and they sponsor the annual *quinceañera* tea, celebrations of US and Mexican holidays, and seasonal parades and beauty pageants. Each year they honour individuals who 'contribute to the cultural development' of the county. In 1995 the commission used its share of tax revenue to begin remodelling their local museum to resemble their ancestral church, Nuestra Señora del Refugio, located in Guerrero Viejo, Mexico, their sister city which also was displaced by the dam. The facade of this church is a symbol that appears in much local art and photography. It was the site of baptisms and weddings of the ancestors of present-day Zapatenses, and the plan for its symbolic reproduction in Zapata is a clear indication of determination to preserve their heritage.

While the county government and the historical commission work hand-in-hand, they are opposed by the alliance of the chamber of commerce and the local newspaper. The relationship between these two groups ranges from antagonistic to nonexistent. The chamber of commerce, as an association of local business people, is the embodiment of capitalistic thought; that is, what is good for business is good for Zapata. They promote fishing in a lake which is now so polluted that people become ill from eating their catch. They solicit retirees to winter in Zapata although indicators suggest that this population strains the social service infrastructure while contributing little to the economy. As bitter as the native Zapatenses are towards outsiders, the chamber of commerce is equally resentful of the insiders. The chamber of commerce director provided this analysis of the situation:

> When the lake originally filled up, they [the federal government] would have come and done quite a lot of work here. And the local people were so bitter at that time that they didn't want it – and they were fools. It's very curious. I do not find any local people that have beautiful houses

> that overlook the lake. Very, very strange, because that's your prime piece of property. They should have been able to see that. But, see, they still don't even care about it. (Annette Turner)

The chamber accuses the historical commission of taking county funds (hotel-motel tax) which they are not entitled to. In fact, this is a justifiable accusation; there is some question as to whether the historical commission is legally entitled to use these funds for projects which do not directly serve tourists, but the county government simply ignores the chamber's protests.

In most US communities the local newspaper is one of the primary bearers of local culture. In Zapata the *Zapata County News* is the voice of the outsider. With one exception, only outsiders provide the editorial material, and the tone is decidedly pro-business and pro-tourism. Any negative aspect about the community, such as water pollution, is glossed over. Any aspect of the town's past which does not fit the trajectory of evolution from the primitive to the progressive is ignored. The paper is, therefore, the primary source for distortion of community history. For example, the publisher told a reporter who wanted to do a story on the 1953 displacement that, yes, she could write about it as long as she did not 'say anything negative because we are trying to get these people to forget the past and to move forward' (Isobel Cañamar). The publisher told me that Zapatenses 'are good people but they haven't learnt how to compete yet' (Charles Post). The paper covers local politics, crime and sports, and native Zapatenses necessarily read it and advertise in it since it is the only paper in town, but they do not endorse it, nor do they accept it as the 'official' news of their town. The newspaper, most businesses that cater to tourists, and the Winter Texans operate within their own cultural and economic milieu in Zapata, an apartheid of sorts, imposed and maintained by the dominant culture of the native Zapatense. The following conversation with the director of the chamber of commerce describes a typical frustrated attempt on the part of an outsider to bridge the gap between the two societies:

J: Well, do you have any occasion to work together, like the county fair or – ?
T: No.
J: Nothing brings you together?
T: No.
J: No overlap of members?
T: No.
J: That's interesting. You'd just think there'd have to be –
T: No. We have tried. The binational meetings that I went to, to get historical tours started in this area, I had to drop it for lack of participation from the historical commission.
[I ask about the Christmas tour of old homes in the county, who puts it together and who publicizes it.]

T: I would help publicize it for them, but they don't want any help. Last year I tried and tried and tried. That's the second time I tried to work with [a certain person on the Zapata County Historical Commission] to get some of their heritage projects and all that stuff in the State of Texas Calendar of Events for some tremendous free publicity. She thinks – she just thinks I'm nuts. I'm serious – she thinks I'm nuts. All this free stuff – tremendous publicity! (Annette Turner)

Testing the strength of social cohesion

First- and second-generation resettlement Zapatenses have managed to maintain social cohesion and community identity even as they have been assimilated into the mainstream economy. But how strong are the bonds that unite these people? They survived resettlement but can they continue to withstand, year after year, the relentless onslaught of the national culture? A quick survey of global trends would suggest that they cannot, yet appearances suggest that they will endure. How can we test the strength or predict the future of the community of Zapata? As it happened, while I was conducting fieldwork in Zapata in 1995–7, the community was faced with a serious crisis which provided a means to examine these concerns. Community crises are good measures of societal strength because they tend to lay bare any weaknesses in the infrastructure. Crises often result in acceleration of social processes already in motion within a community. If native Zapatenses were going to lose control of their community and succumb to outside forces in the near future, the drought that devastated the borderlands from 1994 to 1996 would probably have revealed it. An examination of the crisis response styles of traditional ranchers, the tourist business community and the county government provides the necessary material to test the strength of the community.

Severe droughts occur about every twenty-five to thirty years in the southwestern United States and northern Mexico. A generation is usually considered to be about thirty years, so we can assume that a drought occurs about once every generation. Since this is not often enough for most ranchers to develop and perfect coping strategies based on their own experience with drought, they must rely on knowledge acquired either from outside sources or from family and community memory of the environment. Since most Zapatense ranchers are working land that has been in their families for more than eight generations, they have not only the benefit of access to coping strategies perfected over time but also the reassurance of knowing that others before them have faced and survived similar environmental crises. In the following exchange, a second-generation resettled Zapatense reveals the historical knowledge and long perspective which he draws upon as he goes about managing his family's ranch:

J: When did you start having to get permission to irrigate?
D: 1750. (Armando Díaz)

1750 was the year his ancestors received their land grant from the king of Spain, and those grants included riparian rights to the water of the Río Grande.

Using ancient methods of drought mitigation developed and refined by their ancestors over hundreds of years in pre- and post-Columbian Mexico, in Spain and in the North African desert of the Moors, modern ranchers along the Río Grande cope rather gracefully with drought. They monitor closely the vegetation and water supply in the *monte* (rangeland). At just the right moment, they take blowtorches out onto the cactus strewn *monte* and burn the needles off prickly pear so their cattle can eat the succulent plants. They begin to reduce the size of their herds before the land becomes overgrazed. If necessary they buy water for their cattle, carrying it out to their livestock in portable tanks that they keep for just such occasions. In other words, these ranchers are not taken by surprise but factor water shortage into the long-range planning of their business. As one rancher told me: 'Yes, there is a drought, but this is dry country. It is *always* dry country. Some years are just more dry than others' (Gilbert Hart).

The drought of 1994–6 was severe enough that the federal government provided financial aid to ranchers to help them buy feed for their cattle. Most in Zapata County, however, refused to take advantage of it and, when I asked why, told me that it would be like accepting welfare, i.e., a demeaning thing to do. Is this refusal to accept government aid a perfect example of the pride of the *fronterizo* and their disdain for the federal government? In this case, probably not. Rather, it reflects knowledge of the environment and the long view the ranchers take regarding land management. Temporarily propping up a herd of cattle with imported feed does not solve the problem of drought. If overgrazed at any point, rangeland in this region will probably not recover its beneficial plant life. Only by reducing the size of the herd can the *monte* be protected from overgrazing, so most ranchers took that option rather than feeding their livestock with 'food stamps'.

Since it is taken in stride by most *fronterizos*, drought alone cannot be considered to be a crisis severe enough to test the strength of Zapata. What turned the drought into a devastating crisis was, as often is the case, not nature but the work of humans. The economic crisis in Mexico, and the devaluation of the peso which accompanied it, forced Mexican cattle ranchers to liquidate their herds. When the drought hit Texas, Mexican ranchers were already selling great numbers of cattle to US markets. The market was flooded and cattle prices so low as to hardly be worth the effort to transport them to market. The upshot was that, by 1995, at a time when the best drought strategy was to reduce herd size, Texas ranchers found that they were unable to do so. At that point some ranchers did accept government aid.

The drought was also drying up the Zapata tourist industry. The lake level dropped by more than 40 feet. Tourists who normally came to Falcon Lake to fish or swim along the banks found that they had to walk more than 100 yards through mud to find the shore of the lake. Those who brought boats were unable to get them into the water. The water itself, being greatly reduced in volume, concentrated river pollutants which it normally dispersed, making the lake inadvisable for human use. (The chamber of commerce tried to turn that reduction in volume to an advantage by pointing out that it would be easier to catch large bass in a smaller area!) Motels, restaurants, bait houses, liquor stores and gasoline stations were losing customers. Unlike the ranchers, outsiders who operate the tourist businesses have no indigenous coping strategies to rely on. The strategies they chose to use were typical of those used in other communities in the US: they sought alternative enticements to tourists, called on the state and national government to provide aid, and demanded immediate action from the local government. Their most creative endeavour was to come up with another reason, apart from the lake, to bring tourists into Zapata. In 1995, the chamber of commerce/newspaper coalition concocted the Chupacabra Festival, a food, beer and entertainment festivity capitalizing on the current popularity in the US of a Puerto Rican folk character, the *chupacabra*, a hideous demon that kills livestock by sucking out all of their blood. When wondering aloud how in the world the newspaper and the chamber of commerce came to choose such a theme for its festival, I was told by a local rancher, 'I guess they chose the chupacabra because they knew no Mexican [Mexican-American] would come to the "We're-Sorry-We-Inundated-You Day"' (Isobel Cañamar). Of course, we had just been talking about my research on the dam, so the reference to it did not come out of the blue, but still this is a classic example of the native Zapatense's tendency to use the past to interpret the present. The festival itself was widely advertised but sparsely attended, serving more as a local morale booster for the tourist industry than a real economic boon.

Another coping strategy used by the outsiders was to demand of the county commission that the public boat ramp be extended so people could get their boats into the lake. Ignoring water conditions, they insisted that if the ramp were just long enough the fishermen would return. The response by the county commissioners to this request serves to illustrate the strength and determination of the Zapatenses to preserve their way of life even in the midst of economic crisis: they refused even to address the problem, because it would not benefit their own strongest constituents (native Zapatenses) but would benefit those outsiders who profited from the lake. The chamber of commerce pleaded with the commission but the commissioners turned a deaf ear. Then the chamber implored their state representatives to provide state money, but those officials also were brought into office through the local *patrón* system, so they merely referred the issue back to the county. The chamber then offered to put up the money to have the ramp extended but they were put off. Individuals offered to loan the county the money interest-free but still they refused. For all their weeping and wailing and gnashing of teeth, the tourism faction of the population was consistently ignored. When asked why she thought the county commissioners would not extend the ramp, the chamber director saw it mostly, but not entirely, in economic terms: 'Because I don't think they're real fishermen, and none of our commissioners or our local people in office derive any of their income from tourism. So they tend to ignore it. If it hit them in the pocketbook, they'd be the first ones out there' (Annette Turner).

Finally, during the summer of 1996, one of the last months of the drought, the commissioners agreed to extend the ramp by 20 feet. When asked what took them so long to respond, they claimed that the county simply had not had the money before that time. The real reason was that three years of drought had caused the curtailment of both ranching and tourism income, and the reverberations could be felt throughout the community. Economic losses were pervading all sectors of the county and the politicians were ultimately compelled to respond, if only minimally and too late. In the end, Zapatense businesspersons and the chamber of commerce did join forces to pressure the county to extend the boat ramp. They called a public meeting and press conference where leaders of both insider and outsider factions addressed the need to mitigate the crisis. When asked later to comment on this *ad hoc* alliance, however, Zapatenses would not admit to it. They simply refused to acknowledge the fact that they had worked with their enemy on something. I did get one Zapatense to admit that at times one must do 'what is expedient'. Every Zapatense reverted back to the old antagonisms without explanation. It has already been deleted from community memory.

This unacknowledged and temporary alliance may mark the beginning of a change in social relations in Zapata, but I do not think so.[11] Rather, I believe that it demonstrates the flexible strength of social cohesion which is part of their history there. They bend, taking in occasional outsiders and 'outside' ideas, but they do not break. The community continues to operate with parallel and often antagonistic cultures and economies, and native Zapatenses continue to maintain control over the course of the town. A severe economic crisis did not significantly alter the power relations and only temporarily altered the composition of institutions within the society.

'Official' community history, the dominant culture and politics in Zapata, Texas, are determined by the natives and the natives only. This independence is supported by oil and gas production, which provides great material gains with minimal negative impact on local culture. But this story would not be complete without mentioning the role that the character of these *fronterizos* has played in their own history. The tenacity of these people in clinging to their land and to their culture in the face of a brutal natural environment and a divisive political economy is remarkable. The culture and character of *fronterizos* was forged first in the isolation of a harsh environment and then in conflict with imperialist cultures, and these experiences have led to a greater sense of independence, strong kinship ties and courage to resist outside pressure.

The adversity and opposition encountered by *fronterizos* has shaped their culture, their sense of ethnic identity and the extent to which they will tolerate outside forces. A *gran ofensa* (great offence) is commonly dealt with by refusing to acknowledge that the offence or its perpetrator even exists. A quick survey of things Zapatenses refuse to acknowledge on a daily basis includes the lake in general, anything built by the federal government, tourists, retirees, and any outsiders associated with the other categories. It is this characteristic that makes it fairly easy to see how distortion of their community history and culture is avoided. They shun that which offends them and, so far, they have had the economic and political clout to get away with it. They condemn the villain by refusing, in spite of major obstacles, to allow the villain a place in their world view. This second-generation resettled Zapatense eloquently expresses the prevailing attitude of his people:

> A newspaper man once said to me, 'You are very clannish.' Well, maybe we are. They don't understand how we feel about certain things. Look, when you get a beating, you don't forget. We were able to stick together over here and, yes, it made us different. Because we are the same people who came here in the 1700s. We live here, we die here, we are buried here. We've gone through a lot of droughts and what-have-you, but we are still here. We survived. We are survivors. (Roberto Mendoza)

The prevailing theory in resettlement studies assumes that a resettlement community is a long-term success 'when management of local production systems and the running of the local community are handed over to a second generation that identifies with that community'.[12] Based on this criterion, the resettlement of Zapata can be considered a long-term success. They survived resettlement and, what is more, they have survived the onslaught of US cultural imperialism that followed it. They maintain much of their indigenous past even while accommodating the national political economy. Second-generation Zapatenses not only control the economic and political spheres of their town during good times and bad, they control its 'official' community memory and culture as well.

Notes

1 J. Knippers Black, *Development in Theory and Practice: Bridging the Gap* (Boulder, CO, 1991), vi.

2 M. M. Cernea, *The Risks and Reconstruction Model for Resettling Displaced Populations* (Oxford, 1996), 1.

3 Pseudonyms are used for all of the individual inhabitants of Zapata referred to in this chapter. The quotes are excerpts from interviews I conducted during fieldwork in Zapata County, Texas, US, and in Guerrero Nuevo, Tamaulipas, Mexico, from June 1995 through August 1997. The oral history interviews were funded by the Baylor University Institute for Oral History and are deposited there.

4 The term *Anglo* is used by Mexican-Americans throughout South Texas to refer generally to all light-skinned, non-hispanic people living in the US. In Zapata, they also often call them *Americans* while referring to themselves as *Mexicans*. This is a linguistic distinction based only on appearance and culture, however, and not on national allegiance.

5 The earliest known inhabitants of this region were the Coahuiltecans, a widely scattered foraging society killed off during the colonial period. They were followed by the Apache, the great horse society of the Southern Plains, and the Comanche, both of whom terrorized colonial and nineteenth-century settlers throughout the southwestern US and northern Mexico.

6 E. Spicer, 'Persistent Cultural Systems: A Comparative Study of Identity Systems that Can Adapt to Contrasting Environments', *Science*, 174 (1971), 795–800.

7 W. Roseberry, *Anthropologies and Histories: Essays in Culture, History and Political Economy* (New Brunswick, NJ, 1991).

8 A. M. Alonso, 'The Effects of Truth: Re-presentations of the Past and the Imagining of Community', *Journal of Historical Sociology*, 1(1) (1988), 33–57.

9 N. G. Canclini, *Transforming Modernity: Popular Culture in Mexico* (Austin, TX, 1993).

10 Ibid.

11 I should mention here the only other occasion on which I observed behaviour that suggested a loosening of Zapatense cohesion. In 1996, while attending a meeting of the Zapata County Historical Society, I was surprised to find them discussing the institution of a Protestant religious component to their annual *quinceañera* celebration. The *quinceañera* is a traditional Mexican coming-of-age ritual for a girl on her fifteenth birthday. The usual celebration includes a Catholic mass, but the historical society encourages any girl – Anglo- or Mexican-American, Catholic or Protestant – to participate. Of course, the inclusion of a Protestant component to a traditionally Catholic-tinged cultural event may reflect a trend throughout Latin America, which is being heavily evangelized by Protestants, but I was surprised to find it in Zapata.

12 T. Scudder and E. Colson, 'From Welfare to Development: A Conceptual Framework for the Analysis of Dislocated People', in A. Hansen and A. Oliver-Smith (eds), *Involuntary Resettlement and Migration* (Boulder, CO, 1972), 267–87.

Select bibliography

Alonso, A. M., 'The Effects of Truth: Re-presentations of the Past and the Imagining of Community', *Journal of Historical Sociology*, 1(1) (1988), 33–57.

Black, J. K., *Development in Theory and Practice: Bridging the Gap* (Boulder, CO, 1991).

Cernea, M. M., *The Risks and Reconstruction Model for Resettling Displaced Populations* (Oxford, 1996).

García Canclini, N., *Transforming Modernity: Popular Culture in Mexico* (Austin, TX, 1993).

Roseberry, W., *Anthropologies and Histories: Essays in Culture, History and Political Economy* (New Brunswick, NJ, 1991).

Scudder, T. and Colson, E., 'From Welfare to Development: A Conceptual Framework for the Analysis of Dislocated People', in A. Hansen and A. Oliver-Smith (eds), *Involuntary Resettlement and Migration* (Boulder, CO, 1972), 267–87.

Spicer, E., 'Persistent Cultural Systems: A Comparative Study of Identity Systems that Can Adapt to Contrasting Environments', *Science*, 174 (1971), 795–800.

7 Signs of things to come

Metaphor and environmental consciousness in a Yucatecan community

David W. Forrest

Metaphor is basic to the way humans understand the world in which they live. For the Maya, who have a tradition of conceptualizing time as cyclical, the understanding of current events in terms of the past and future is a well-established way of understanding. This article explores how the people of the contemporary Yucatecan Maya community of Maní in south-eastern Mexico use metaphor to help in the construction of environmental consciousness by linking current events to those of the past and those prophesied for the future.

Maní, a town of approximately 3,500 people, is located 70 kilometres south of the state of Yucatán's capital of Mérida to the north of a range of low hills known as the Puuc. The northern part of the Yucatán Peninsula is a low-lying limestone shelf characterized by a tropical wet-and-dry climate, with a wet-season of abundant rainfall from May to October and a dry-season of low rainfall from November to April. Maní receives approximately 1,200 millimetres of rain per year, with a low of about 25 millimetres in February and a high of about 219 millimetres in September.[1] There are no surface streams or lakes in this area and much of this rainfall seeps through the porous limestone where it recharges the groundwater. This downward percolation of acidic rainwater has created a karst topography characterized by pock-marked limestone, sinkholes and caverns. Most of the water traditionally used by the population is either from natural sinkholes that lead to subterranean sources of water, locally known as *cenotes*, or from human-made wells leading to the aquifer.

In response to living in an environment characterized by a long dry-season and a lack of surface water for irrigation, the ancient Maya of Yucatán developed a two-part infield-outfield agricultural system. With a few modifications, this system continues to be used in many parts of Yucatán today. In this system, the infield is represented by the *solar*, a traditional form of the homegarden which is found in many areas of the tropics.[2] Within the *solar*, which is enclosed by a dry-laid stone fence, are the household's fruit trees, medicinal and culinary herbs, pigs, chickens and turkeys. Maya homegardens are highly diverse in terms of species composition and highly productive for their small size. Many of the plants within the homegarden require irrigation from the household's well to survive and produce, especially during the long dry-season. Some favoured plants include various types of citrus (such as oranges, limes and tangerines) and several varieties of bananas and plantains which were introduced by the Spanish but adopted by the Maya and given Maya names. As with homegardens in many parts of the world, the *solar* is under the control of the women of the household.

The outfield is represented by a rain-fed swidden plot, called *milpa* in Spanish or *kol* in Maya. The *milpa* is often cut from forest in the *ejido*, the lands held communally by the town, where the *milpero* (*milpa* farmer) has usufruct rights to the plot while it is under cultivation. The *milpa* was traditionally used to produce annual staple crops requiring high soil fertility and sunlight, together with small amounts of other vegetables.

The role of *milpero* is the traditional role of Maya peasant males (*campesinos*) in Maní. After undertaking prescribed rituals and making offerings to the *Yuntziloob*, the spirit guardians of the forest, the *milpero* cuts the vegetation of the plot (today about 2 hectares in size) early in the dry-season (about December–January). He allows this vegetation to dry thoroughly for several months and then he burns the plot before the arrival of the rainy-season (about May). Soon after the plot is burnt, seeds are planted using a metal-tipped digging stick known as a *xu'ul*. Traditionally, maize, beans and squashes were planted as staple crops for the household, along with smaller amounts of other food crops such as watermelons. The local varieties of maize used in the *milpa* are genetically variable and more drought and pest resistant than hybrid varieties. This plot was usually used for two years, then abandoned and allowed to lie fallow for a minimum of eight to twelve years (preferably twenty years) before being used again. Until recently, this system characterized the economies of most households in Maní. Indeed for most of Maní's history, the fate of its people has been tied to production in the *milpa*.

Although today Maní appears to the outsider as a Maya town not unlike others of the region, it holds a special place both in Maya cosmology and in the history of the peninsula. At the time of the arrival of the Spanish in Yucatán in the sixteenth century, Maní was the largest town in the region. Early Spanish explorers reported more than 1,000 houses in the town, representing a population of approximately 4,000 people.[3] As the administrative seat of the dominant Xiu lineage, Maní was a political and cultural centre for the peninsula. During the first two centuries of the colonial period, the population of Maní and the rest of the peninsula declined drastically. This was due largely to the introduction of European diseases such as smallpox to which the indigenous population was highly vulnerable. It is only in the late twentieth century, with a population of about 3,500, that Maní is again approaching the population level that it held at the time of the Conquest. Although Maní has lost most of its political and economic importance over the past five centuries, it has retained much of its cultural importance to the Maya of the region as the subject of many important Maya myths.

The Maya of Yucatán, along with many Mesoamerican societies, hold a sense of time that is both cyclical and pessimistic. Maya notions of imminent environmental collapse are widely known and have been explored by several researchers in recent years.[4] This world view may in fact be the result of hard-won lessons of the distant past. Current scholarly thought holds that ecological collapse, as a result of a drying climate which exacerbated the stress on an already overexploited environment, was a major component in the downfall of Classic Maya civilization in the lowlands which began around 800 AD.[5] Interruption of the region's food supply has been a recurrent problem in Yucatán throughout its history. During the colonial period for example, Yucatán suffered at least nineteen major natural disruptions in the food supply leading to famines.[6] Such natural disasters leading to food shortages continued throughout the nineteenth century.[7]

In contemporary Yucatecan communities, local ecological history is recorded in the narratives that people tell about the environment. It is through these narratives that history is brought into the present day. As intrinsic parts of the landscapes of these communities, narratives provide a medium for the expression of the community's metaphors about the nature of the environment and the relationship of humans to the resources found there. These metaphorical links between the past and the present became clear to me while I conducted fieldwork in Maní in 1994 and 1995. One of the methods of data collection that I used was a type of walk-as-interview that I came to call *rumbos* (Spanish for a road, route or way). On these walks I travelled with people of different ages and experiences along routes (paths, roads and streets) that they knew well to make use of the places and natural resources we encountered along these routes, to elicit narratives about the environment.[8] Many of these stories I might not have been told, nor would I have known to ask about, had we conducted the interviews without the particular landscape of the route as a stimulus. I wrote down notes on a note pad as we walked, then filled in the complete narratives during pauses we made along the route, often getting clarifications on wording from my companion at that time. I was amazed at the richness of the narratives told to me on these walks. In them people mixed together talk about the everyday use and control of resources with talk about the community's past and its place in the cosmos. People often talked about current events by referring to the past, talked about history by referring to mythology, and talked about local geography by referring to Maya cosmology. By linking environmental conditions of the past to those of the present, these metaphors serve an important role in the creation of local environmental consciousness.

Reflecting the cycles of forest transformation into the *milpa* and its regeneration as forest, along with larger cycles of environmental collapse and recovery, is the Maya concept of cyclical time. The calendar of the ancient Maya was composed of a number of nested and interrelated cycles. The *katun* round (*u kahlay katunoob*) is a repeating cycle of thirteen twenty-year periods called *katunoob*, each with its designation and characteristic events. The *katun* count is based on the 360-day *tun*, rather than the 365-day solar year (*haab*).[9] The same pattern of events for a particular *katun* is expected to occur again when that *katun* is repeated 260 *tun* years (265 solar years) later. The ending of any cycle in the Maya calendar was seen as a time of inherent uncertainty, and the end of major cycles was accompanied by dread of the approaching end of the world. Although the modern-day Maya of Yucatán no longer preserve the day and month names of the ancient calendar, they do preserve the concepts informing it. The yearly almanac represented by the *xoc k'in* (the count of days), in which the weather of the first twelve days of January is used as a prophecy of the ecological conditions of the coming year, is part of every traditional Maya *milpa* farmer's repertory of knowledge about the environment. Dionisio Can Ek[10] of Maní described the *xoc k'in* to me in 1994.

> *Los Abuelos* [the ancestors] said that the month of January serves as the almanac for the whole coming year. The weather on each day of January will foretell the weather for the months of that year. The weather on the first of January is what will happen in January, the second of January is what will happen in February, the third of January is what will happen in March. They could tell by what happened on these days of January how the rains would fall for the year. This is called the *xoc k'in*. In this period called *xoc k'in* they could tell from the movement of the winds and the rains what would happen. By using the *xoc k'in* they could tell what day one should do certain things. *Los Abuelos* were wiser about the times for rain than are the people now.

With a cyclical notion of time, events occurring in a particular cycle are seen as representative of that cycle, and are expected to be repeated in the following cycle.[11] The Yucatecan literature known as the *Books of Chilam Balam*,[12] which are collections of *katun* counts, prophecies and stories that were written down in European script after the Spanish Conquest in the sixteenth century, are post-Conquest versions of ancient Maya texts.[13] One of these, the *Book of Chilam Balam of Maní*,[14] contains the prophecies made by a *chilam*, or priest, who lived in Maní during the late pre-Conquest period. Among these prophecies are echoes of the uncertain nature of local food production made worse by the occurrence of natural disasters. One type of disaster common throughout Yucatán's past is plagues of locusts. One such plague occurred in the sixteenth century and is described in the *Book of Chilam Balam of Maní*.

> *Kinchil Coba* will establish a *Katun 13 Ahau*. The stone on which it will be written will be in Mayapán. *Itzamna, Itzamna-tzab* will be its countenance. There shall be much famine. For five years bread made of the *cup*[15] shall be eaten because there will be three years when a plague of locusts will devour the plants and flowers and lay their larvae by the millions. The ruler will have his eyes on the heavens and the stars, and there shall be eclipses of the sun and the moon.

A similar event occurred in the 1940s in Maní and was recounted for me by Dionisio Can Ek while I walked with him to his *milpa* one day in 1994.

> In 1942, 1943 and 1944 there were plagues of locusts three years in a row. They would come in September, stay until they ate everything, then disappear; and then come back the next year. They ate everything, even all the leaves on the trees.
>
> We [the family] ate *ramón* fruits, we made tortillas out of them; when all these things were eaten, we began to eat sour oranges. Some people ate the heart of the *k'uun che'* [*Jacaratia mexicana*]. They took off the outside part and mashed and milled the soft interior part and made *tortillas* to eat; but the people who ate only this died.[17]

Separated by over four centuries, these accounts, one of which occurred in the sixteenth century and the other in the 1940s, mirror each other in their duration (three years) and their consequences. Although the account from the *Book of Chilam Balam* is a prophecy, telling of the future, and Don Dionisio's story is local history, telling of the past, they both make metaphorical connections between the vicissitudes of *milpa* production in the here and now and similar uncertainties in the beyond-the-everyday realm of the past and future. Yucatán was and is subject to a number of natural phenomena which can destroy the *milpa* harvest of entire regions. Tropical storms and droughts can destroy the maize crop for entire communities. Plagues of locusts often accompanied droughts in the past, compounding the problem of food scarcity. Single crop failures occur frequently; it is the occurrence of several successive years of crop failures, such as described in the *Book of Chilam Balam of Maní* and by Don Dionisio, that causes true famines.

The locust plagues of the 1940s were a major disruption in the life of the community of Maní. Many people were forced to leave the community in search of wage labour to buy food for their families. Many of the others who were not able to leave the town died of starvation. As a result, the 1950 population of the municipality of Maní, six years after the plagues had ended, was still about 12 per cent lower than what the population size had been in 1940.[18] The memory of these events is still strong in Maní.

> After the locust plague [1942–4] it was six to eight years before the harvest of maize began again in Maní. At first men went to other places to look for work, then sent for their families. Many went to Santa Rosa on the route to Peto or to Catmis, an *hacienda* with a lot of sugar.[19] Many suffered and died. The men who went away returned each month to see their families. (Dionisio Can Ek)

> Martín Quijada's father and uncle went to Catmis and another *hacienda* called Santa Rosa near Peto to work during the locust plague. (Beatriz Guerrero Méndez)

Environmental crises and the prophecies about them are not only a thing of the distant past. Prophets still appear in Maní to tell of the signs of things to come. Their narratives represent present-day versions of the Maya sacred books of the past. In 1994, Don Dionisio told me about such a prophecy that was made in the 1960s.

> There came to the park [in the centre of town] a prophet to tell of signs of things to come. He said that in 1970, the people would have suffering but not much. In 1980 and 1985 there would be wars. In 1995 there would be people with food but not contentment, more difficult things. *Chichnak*, discontented people . . .

Foretold over thirty years ago, the agricultural crisis of 1995 was in some ways a new type of environmental crisis for Maní. It was caused not by a drought, a hurricane or a plague of locusts, but by the inability to purchase the agricultural chemicals needed to produce cash crops on which many of the community's households now depend.

In Maní and other communities along the Puuc Hills, government sponsored irrigation projects, known locally as *parcelas*, have been established during the last thirty years. Sections of the town's *ejido* land (land for which usufruct rights are granted to community members) have been used to create irrigation units, each divided into between thirty and fifty parcels of 1–3 hectares in size, and supplied with a well, an electric pump and the necessary canals, pipes and tubes to carry water to each parcel. In Maní the first of these irrigation pumps began functioning in 1961. By 1994, eight of these irrigation units were functioning in the community and four more were due to be completed in 1995 and 1996. The *parcelas* are linked to Maní's past through the present-day narratives about the founding of Oxkutzcab, a nearby town to the south of Maní.

> When the Xiu and their soldiers began to live here [in Maní] they began to look for lands to make *milpa*, then afterwards they began to look for better lands to make their vegetable plots. They then went to the south near the hills [Puuc] and first they planted maize; then they planted *ramón*, *ox*;[20] next they planted tobacco, in Maya *kutz*; then they made hollow tree trunks for bees, *cab*. (Dionisio Can Ek)

This story of the founding of Oxkutzcab and the gardens of the Xiu is an example of how the past can be reinterpreted as prophecy. This link between current land use and events of the past both informs the understanding of the past by drawing on knowledge of the importance of *parcelas* near Oxkutzcab today, and alludes to the prophetic nature of this name by viewing it as a prediction of the development of the *parcelas* of the late twentieth century.

The *parcelas* have allowed a change from rain-fed shifting cultivation to long-term production on permanently tenured irrigated parcels, a major shift in the type of agricultural production for Maní and other communities in the region. This new technology freed agricultural producers in Maní from a reliance on rain-fed agriculture on a large scale for the first time in the history of the community. As one man in Maní told me,

> When I was a boy, the rains used to start on the sixth of July, but now they don't start then. There is less rain than before, but it is different. People have *parcelas* now and with irrigation and fertilizers they can have a harvest. (Francisco Gálvez)

Some parcels have a wide variety of fruit trees and horticultural crops, such as mangoes, various types of citrus, bananas and plantains, coconuts and various local fruit varieties, echoing the diversity of the traditional *solar*. Other parcels, especially those in the newer irrigation projects, are planted almost entirely with citrus species. Citrus, especially varieties of oranges used for juice, is in great demand in the restaurants of tourist resorts along the Caribbean coast such as Cancún and Cozumel. The market for local oranges is further increased by the location nearby of an orange juice concentrate processing plant which ships this concentrate to other parts of Mexico and abroad.

Some people in the community have resisted this trend towards more intensive, market-orientated agriculture. Through their narratives, they express the concern that by turning the *ejido* lands of the community into *parcelas*, there will be future shortages of land for staple food production.

> Twenty years ago there was more maize, since there were no *parcelas*. There were 1–2 hectares of *milpa* for each person, some had 4 hectares of *milpa*, mostly maize and beans. In good years, the harvest of maize could be double. (Cecilio Cocom Pat)

> Before [the *parcelas*] in Maní, people raised almost all their food. They were self-sufficient. They only bought flour and sugar, the rest they grew here. (Andrés Parada May)

> When the *parcelas* began in Maní, there was a man whose family was very involved in politics that didn't want there to be *parcelas*. He said, 'Where are the *campesinos* going to find *monte* [forest] to cultivate?' Now in Maní there is hardly any *monte* left to plant maize, the *parcelas* have it all. You have to rent land from a private property to plant maize. Now, this man who didn't want *parcelas*, his sons now have two or three *parcelas* each. (Francisco Gálvez)

Falling market prices for maize and the declining productivity of *milpas* in overused *ejido* lands on the one hand, and the opening of new markets for locally produced fruits and vegetables on the other, have pushed many people into more intensive agricultural production. In the case of the *parcelas*, this has led to economic dependence on irrigated agricultural production and an increased dependence on access to water provided by the government for a source of income. Even those who do not own parcels, such as day-labourers, depend on income from agricultural work in the irrigated parcels. For many in Maní, the *parcelas* have become synonymous with work and income; a sentiment which is commonly voiced in their narratives.

> Before the *parcelas* there was no work for wages in Maní, all around there wasn't any wage work. Only the owners of *ranchos* paid people to work to cut *monte* [forest] to make *milpa* for them. (Cecilio Cocom Pat)

> I started going to work in the *milpas* about six years ago [at nine years old] with my *coa* [small hooked knife] to use to cut, but I got tired of it and started going to work in the *parcelas*. That is almost all I do now. (Miguel Carrillo González)

In addition to the irrigated *parcelas*, vegetables produced for local and regional markets are replacing maize in the *milpas* around the town. The use of chemical herbicides, insecticides and fertilizers has increased in the community with the increased production of vegetables on these more intensively cultivated plots. About 45 per cent of the households in the community with *milpas* reported that they used chemicals on their *milpas* in 1994.[21]

This trend is resisted by more traditional *milpa* farmers in Maní who still produce maize for subsistence as their primary occupation. These producers espouse metaphors of resource use in which the community's resources are seen as a source of sustenance. The forest is referred to as a sacred realm, guarded by the *Yuntziloob*, supernatural forces that require offerings from the humans who temporarily use plots on which to grow maize to provide for their sustenance. Small-scale offerings and ceremonies are performed by the *milpero* and his family at various times, such as before cutting the forest for a *milpa* or on reaping the first fruits of the harvest. Many *milperos* have a small *ka'anche'* or *mesa* (sacrificial altar) in their *milpa* for this purpose. Other ceremonies are more elaborate and require the services of a *hmen* (shaman), such as the *Chachaac* (rain ceremony) or the *han li kol* which are performed to placate the *Yuntziloob*. In this traditional view, people and the forest are seen as part of a set of interrelated cycles of growth and decline, drawing on centuries of Maya practice and belief. The narratives of many Manienses exemplify this traditional view.

> [While walking along a path with low-hanging branches over it] This path is easy for short people, not tall people. It is made for *aluxes* [dwarf-tricksters who inhabit the forest]. (Miguel Carrillo González)

For the *primicia* [offering] for your *milpa*, you take five *jícaras* [calabash bowls] of *sac ha* [ground maize, water and honey] and place them on a *ka'anche'* to offer them as a *primicia* for the *milpa*. The first *primicia* you do when you measure the *milpa* before you begin to cut it, then also at times afterwards. (Antonio Hunac Ceel)

Some people always ask permission before cutting the forest for a *milpa*. You do this with *sac ha*. If you don't, you will meet with a lot of snakes in the plot. Some people like the Ucan never enter a forest to cut a *milpa* without first making an offering of *sac ha*. They say they do this and then they don't see a snake the whole time until the harvest. (Dionisio Can Ek)

The *Chachaac* is for rain. There are other ones [rituals] that they do: *Sac ha* is with *sac ha* in *jícaras*, it is done for spirits. The *han li kol* is for spirits and is done in the *milpa*, the *rancho*, the *solar* . . . (Miguel Carrillo González)

These traditional metaphors for community resource use have been pushed aside by many community members who primarily grow vegetables and fruits as cash crops, and then purchase most of the maize their families consume. These producers embrace metaphors of resource use in which the community's resources are referred to as a source of wealth and a means through which they can obtain a better life for themselves and their families. The forest is seen as exploitable land, which with inputs of capital and labour can produce a profit in a linear move towards progress. This new relationship between Maní's people and resources is expressed in the narratives of these market-orientated producers.

It is possible to make good money from a *milpa*. There are people here in Maní who make a *milpa* and plant only watermelon. From 1 hectare of only watermelon you can get at least 10 tons and up to 12 or more tons, and you can sell it for 8,000 pesos. One hectare *milpa* of maize gives 1 ton that you can sell for 800 pesos, very cheap. There are expenses though with watermelon. You have to spend 1,500 pesos to 2,000 pesos on fertilizers and herbicides that you have to use to get a good harvest. But still you make 6,000 pesos,[22] and then you can buy maize to eat. The people who do this have very nice houses and pickup trucks, because they can buy things after the harvest. (Pedro Herrera Chi)

This is the property of a man from Oxkutzcab . . . He has all this planted in lemons, all the way to those coconuts; and over there all grapefruit. There is also squash and cucumber in the first part with the lemons . . . They say he sells his lemons directly to people in other places, trailers come and take them away. He has so many that there are too many to sell in Oxkutzcab. (Juan Castro Morales)

Amendments to Article 27 of the Mexican Constitution made in 1992 marked the end of the *ejido* system of communal land tenure and launched a programme to distribute the *ejido* lands to *ejido* members as permanently tenured plots. These legal changes will most likely further the trend towards more intensive methods of agriculture. As the shifting agricultural system of the past disappears in favour of permanently cultivated plots, the land around the town is being shifted towards more intensive forms of resource exploitation. The income generated from these new forms of production has become the primary mechanism for the growth of Maní in the past few decades. As one man told me in 1994,

> All the area around Candelaria chapel that is now built up was only vacant lots until twenty to twenty-five years ago. The stores there are new. The *parcelas* caused all this growth. Now there is not as much unemployment in Maní as there was before.
>
> With these *parcelas* the people of Maní have their own way of life. In all parts of Maní there are new houses, some of them big houses. This is a result of this new way of life in Maní. There are more people now here in Maní than before. (Francisco Gálvez)

Although much of the growth occurring today is caused by the rise of these more intensive forms of agriculture, people know that this type of growth has happened before. This knowledge is held in the narratives about the community's history. The town's recent surge of construction and increase in population is linked to this history by narratives about Maní in the distant past, before smallpox and plagues of locusts reduced and dispersed the town's population. As Dionisio Can Ek told me,

> *Los Abuelos* [the ancestors] said that Maní used to have many more people than now. All of these lots that are forest now used to have houses and families living in them. But then came the plague, how do you say it, the smallpox and then later the plagues of locusts and many people died and went away, and the *solares* were abandoned. But it is going to fill up again like before. You can see it happening now . . . Like the story in the Bible: After the Flood there sprouted the seed of Noah. It was like that in Maní, after the plague and the locusts many people had gone, but still the seed was there to come up again.

Events in 1995, however, called into question the permanence of this 'new way of life in Maní'. The Mexican economic crisis of 1995 exposed the dangers of the increased dependence by many local producers on a few or even one intensively cultivated crop such as oranges or watermelons.

> The price for oranges and mangoes is the same as last year [1994], but the chemicals have gone up a lot [twice the price] . . . Many people can't

afford to buy them. It is not too bad for oranges, you don't spray as much. But with cucumber and watermelon that require a lot of spraying, it's a big problem. (Cecilio Cocom Pat)

As a result of the 1995 peso devaluation, the prices for agricultural chemicals used in Maní doubled from their 1994 prices. At the same time the prices for Maní's produce in local markets remained approximately the same as the previous year, presenting a major problem for producers who could not afford to buy the chemicals needed for their production. The 1995 crisis resulted in many 'discontented people' and so, in the minds of many, fulfilled the prediction spoken years earlier by the prophet who came to the park in the centre of town.

The events of 1995 and the 1940s, and those of the more distant past recorded in the narratives and prophecies of Maní, are understood as similar events that have occurred at different times but with similar results. In both the 1940s and the 1990s, the migration of the town's people was not into the forests in search of food as it had been during the colonial period, but into the market economy. In the 1940s people fled to vast sugar and *henequén*[23] *haciendas* in other parts of Yucatán to seek agricultural work, while in 1995 they moved into the urban and tourist resort areas of the peninsula in search of wage labour.

By linking present-day trends in resource use to periods of growth and decline in the past, the *parcelas* and vegetable plots are made to seem a part of the history of the community, but also part of the long-term cycles of growth and decline. Intensive cultivation on the *parcelas* and vegetable plots is the mechanism for much of Maní's recent growth, but by looking at past cycles as known through narratives, people expect that this growth will turn to decline once again in the future. In 1994 Dionisio Can Ek told me about the humble future that had been foretold for Maní many years ago.

> *Los Abuelos* [the ancestors] also said that Maní is not going to grow much; that Maní is not going to become refined; Maní is not going to be large, it is going to remain like this because *los Abuelos* said so. Those that want to build won't be able. They will still live, but they won't become rich. For this reason, Maní only has a little change.
>
> Before there were a few masonry houses, the rest of Maní was all Maya houses [*casas de paja*] . . . The *parcelas* began to grow twenty to twenty-five years ago in Maní, and then people began to build masonry houses in Maní.
>
> But still there is only a little change in Maní, like the Pérez, who sell things. But in Dzan [a nearby town], there are many things. Many things are sold there, but not in Maní; pure working the *milpas* up until now.
>
> Already many of these things have come true. It is true what *los Abuelos* said. In their time, no one knew how to read, but they knew these things by listening to their forefathers and telling it again.

By linking past and future environmental conditions to those of the present, metaphor is used to defend and advocate types of resource use and control that currently exist in the community. Through their narratives, traditional *milpa* farmers like Don Dionisio on the one hand, and more market-orientated producers on the other, contest discourses about the best use of the land and water resources of the community. For the most part, the capitalist metaphors seem to be gaining strength in Maní as each year the town's producers become more involved in the production of fruits and vegetables for the urban market. In the economic crisis of 1995, however, many of the households that depend on income from fruit and vegetable production to purchase staple foods went hungry as their profits dropped. For the first time in many years, the metaphors of the traditional *milpa* farmers received an affirmation as they were brought out in a positive contrast to those of the more market-orientated producers. In addition, prophecies of the past were realized and the adoption of new methods of agriculture came into question as people were forced to re-evaluate current events in terms of the environmental crises of the past.

E. N. Anderson has argued that the real challenge in the conservation of natural resources is not in managing the resources themselves, but in managing and motivating the people who use the resources.[24] In Maní, metaphor serves as a powerful tool which aids in this process by helping the people of the community create a context for understanding their environment. By bringing the knowledge of centuries of local environmental history into the present, narratives about the environment help people to create an environmental consciousness which remembers past environmental crises and anticipates those of the future. In a community whose people expect there to be environmental collapse, referring to events of environmental crisis dispersed throughout time is a powerful means of understanding the possibilities and limitations of their environment in the present.

Notes

This chapter is based on research that I conducted in Maní in 1994 and 1995. I wish to thank Jaclyn Jeffrey for her comments on an earlier draft of this article.

1 Comisión Nacional de Agua, 'Datos de precipitación, 1973–1992, Distrito de Ticul', unpublished report (1994).

2 E. C. M. Fernandes and P. K. R. Nair, 'An Evaluation of the Structure and Function of Tropical Homegardens', *Agricultural Systems*, 21 (1986), 279–310.

3 S. Cook and W. Borah, *Essays in Population History: Mexico and the Caribbean*, vol. II (Berkeley, CA, 1974), 35.

4 A. Burns, 'Spoken History in an Oral Community: Listening to Mayan Narratives', *International Journal of Oral History*, 9(2) (1988), 99–113; A. Burns, 'There Will Come a Day: Present-day Mayan Thoughts about Ecological Collapse', paper presented at the Institute of Maya Studies, Miami, Florida, 12 October 1990; P. R. Sullivan, 'Contemporary Maya Apocalyptic Prophesy: The Ethnographic and Historical Context', unpublished PhD thesis (Johns Hopkins University, 1984).

5 D. A. Hodell, J. A. Curtis, and M. Brenner, 'Possible Role of Climate in the Collapse of Classic Maya Civilization', *Nature*, 375 (1995), 391–4; J. A. Sabloff, 'Drought and Decline', *Nature*, 375 (1995), 357.

6 These colonial period famines occurred in 1535–41 (drought, locusts), 1564, 1571–2, 1575–6 (drought), 1604, 1618 (locusts), 1627–31 (storm, locusts), 1650–3 (drought), 1692–3 (hurricane, locusts), 1700, 1726–7, 1730, 1742, 1765–8, 1769–74 (drought, locusts, hurricane), 1795, 1800–4 (drought, locusts), 1807 (hurricane) and 1809–10. N. M. Farriss, *Maya Society under Colonial Rule* (Princeton, NJ, 1984), 61–2.

7 Such events occurred in 1818 (storm), 1822–3, 1834–5, 1836, 1842, 1844 (drought), 1854 (drought, locusts) and 1855 (locusts). P. Bracamonte y Sosa, *La memoria enclaustrada: historia indígena de Yucatán, 1789–1860* (Mexico City, 1994), 66.

8 All of the *rumbos* were conducted with male informants ranging from 12 to 69 years of age. Local propriety prevented my taking extended walks with female informants.

9 Farriss, N. M., 'Remembering the Future, Anticipating the Past: History, Time and Cosmology among the Maya of Yucatán', *Comparative Studies in Society and History*, 29(3) (1987), 570.

10 This and the names of other Manienses used in this article are pseudonyms.

11 Farriss, 'Remembering the Future'.

12 E. Craine and R. Reindorp, *The Codex Pérez and the Book of Chilam Balam of Maní* (Norman, OK, 1979); M. S. Edmonson, *The Ancient Future of the Itza: The Book of Chilam Balam of Tizimin* (Austin, TX, 1982); M. Edmonson, *Heaven Born Mérida and Its Destiny: The Book of Chilam Balam of Chumayel* (Austin, TX, 1986).

13 V. R. Bricker, 'The Last Gasp of Maya Hieroglyphic Writing in the Books of Chilam Balam of Chumayel and Chan Kan', in W. Hands and D. Rice (eds), *Word and Image in Maya Culture* (Salt Lake City, UT, 1989).

14 Craine and Reindorp, *Book of Chilam Balam of Maní.*

15 The *cup* refers to the fruits of the tree *Brosimum alicastrum*, today called *ramón* in Spanish or *ox* in Maya. These fruits are usually eaten only by animals, but are eaten by humans in times of famine.

17 This story echoes an account, written down in the 1560s by Fray Diego de Landa, of events that occurred after the initial wars of the Conquest and a drought in the sixteenth century: 'They [the Maya] suffered from a great famine, so much that they began to eat the bark of trees, especially of one that they call *cumche'* [*k'uun che'*], which is soft and bland on the inside.' D. Landa, *Relación de las cosas de Yucatán* (Mérida, Yucatán, 1992), 32.

18 In 1940, the municipality of Maní, including the towns of Maní and Tipikal and the surrounding rural estates, had a population of 2,196. Censo Nacional 1940, *Sexta censo de población, 1940: Estado de Yucatán* (Mexico City, 1943). In 1950 the population of the municipality had dropped to 1,945. S. Rodríguez Losa, *Geografía Política de Yucatán*, vol. III (Mérida, Yucatán, 1991), 250.

19 Catmis, 70 kilometres, and Santa Rosa, 80 kilometres south-east of Maní, were major sugar-producing estates during this period.

20 See note 15.

21 These data were obtained from a random survey of fifty-one households in Maní that I conducted in 1994.

22 To put this into perspective, 6,000 pesos in 1994 represented the earnings from 500 days of work by an experienced day-labourer, typically paid about 12 pesos per day.

23 A native plant (*Agave fourcroydes*) cultivated for the hard fibre of its leaves.

24 E. Anderson, *Ecologies of the Heart: Emotion, Belief, and the Environment* (Oxford, 1996).

Select bibliography

Anderson, E. N., *Ecologies of the Heart: Emotion, Belief, and the Environment* (Oxford, 1996).

Bracamonte y Sosa, P., *La memoria enclaustrada: historia indígina de Yucatán, 1789–1860* (Mexico City, 1994).

Bricker, V. R., 'The Last Gasp of Maya Hieroglyphic Writing in the Books of Chilam Balam of Chumayel and Chan Kan', in W. Hanks and D. Rice (eds), *Word and Image in Maya Culture* (Salt Lake City, UT, 1989), 39–50.

Burns, A. F., 'Spoken History in an Oral Community: Listening to Mayan Narratives', *International Journal of Oral History*, 9(2) (1988), 99–113.

—— 'There Will Come a Day: Present-day Mayan Thoughts about Ecological Collapse', paper presented at the Institute of Maya Studies, Miami, Florida, 12 October, 1990.

Censo Nacional 1940, *Sexta censo de población, 1940: Estado de Yucatán* (Mexico City, 1943).

Comisión Nacional de Agua, 'Datos de precipitación, 1973–1992, Distrito de Ticul', unpublished report (1994).

Cook, S. and Borah, W., *Essays in Population History: Mexico and the Caribbean*, vol. II (Berkeley, CA, 1974).

Craine, E. and Reindorp, R., *The Codex Pérez and the Book of Chilam Balam of Maní* (Norman, OK, 1979).

Edmonson, M. S., *The Ancient Future of the Itza: The Book of Chilam Balam of Tizimin* (Austin, TX, 1982).

—— *Heaven Born Mérida and Its Destiny: The Book of Chilam Balam of Chumayel* (Austin, TX, 1986).

Farriss, N. M., *Maya Society under Colonial Rule* (Princeton, NJ, 1984).

—— 'Remembering the Future, Anticipating the Past: History, Time and Cosmology among the Maya of Yucatán', *Comparative Studies in Society and History*, 29(3) (1987), 566–93.

Fernandes, E. C. M. and Nair, P. K. R., 'An Evaluation of the Structure and Function of Tropical Homegardens', *Agricultural Systems*, 21 (1986), 279–310.

Hodell, D. A., Curtis, J. A. and Brenner, M., 'Possible Role of Climate in the Collapse of Classic Maya Civilization', *Nature* 375 (1995), 391–4.

Landa, D., *Relación de las cosas de Yucatán* (Mérida, Yucatán, 1992).

Rodríguez Losa, S., *Geografía política de Yucatán*, vol. III (Mérida, Yucatán, 1991).

Sabloff, J. A., 'Drought and Decline', *Nature*, 375 (1995), 357.

Sullivan, P. R., 'Contemporary Maya Apocalyptic Prophesy: The Ethnographic and Historical Context', unpublished PhD thesis (Johns Hopkins University, 1984).

8 The environmental movement in Kazakstan

Ecology, democracy and nationalism

Timothy Edmunds

Kazakstan is an important example of a former Soviet republic attempting to democratize in a difficult economic, political, ecological and ethnic environment. Containing over 100 different ethnic groups and including a sizeable and regionally concentrated Russian diaspora, no single 'nationality' forms a majority of the population. With reference points such as Yugoslavia or the Caucasus, many western analysts predicted that the post-independence political climate in Kazakstan would be dominated by nationalism and nationalist movements. In reality, however, political and nationalist dissatisfaction in Kazakstan have expressed themselves through other issue areas, most notably environmentalism. Consequently, the role environmental movements have played in the political process in Kazakstan is extremely significant, and remains perhaps unique in the whole of the former-Soviet Union.

Environmental organizations have had a significant impact on the development of a participative culture and 'civil society' in Kazakstan. During the Soviet period, environmental protest was often a symptom of Republic-level discontent over the nature of Kazakstan's relationship with Moscow, and as such became closely linked to pro-Kazak political sentiment. Similarly, a close affinity with the Kazakstani environment was seen to be central to the Kazak 'national' character, further strengthening the relationship between ecology and Kazak nationalism. Today, at a period in time when all forms of political opposition in the Republic are becoming increasingly politically marginalized, environmentalism appears to remain high on the political agenda. Though many environmental groups have suffered from the problems afflicting all opposition groups in Kazakstan, their influence remains significant and nationwide awareness of environmental problems plays a major role in many different aspects of the Kazakstani political debate. For many Kazakstanis, the activity of environmental groups represents the only area of the Republic's political life where it is possible to comment on and influence policy. My intention here is to trace the history of environmental politics in Kazakstan, and analyse its role in the present-day political life of the Republic.

Kazakstan faces a huge variety of environmental problems, a legacy from the Soviet period. The most devastating of these are the desiccation of the Aral

Sea, and the presence on Kazakstani soil of the Soviet Union's primary nuclear test facility: the 'Polygon'.

The decline of the Aral Sea represents an ecological and human tragedy of enormous proportions, directly affecting both Kazakstan and the neighbouring republic of Uzbekistan, and with implications for the whole of Central Asia. The Aral was once the fourth largest body of inland water in the world. Fed by its two tributary rivers, the Amu-Darya and Syr-Darya, the Aral Sea boasted a huge variety of often unique wildlife, and supported a flourishing fishing industry large enough to supply a tenth of the entire Soviet catch.[1] Since the 1960s, however, the Aral has been drying up. Increasing amounts of water from both the Amu-Darya and Syr-Darya have been diverted for agricultural irrigation, to the extent that between 1974 and 1986 the Syr-Darya could not even reach the sea. What water has found its way to the Aral has often been heavily contaminated with agricultural pollutants, including herbicides, pesticides and chemical fertilizers, from intensive cotton farming upstream. To make matters worse, and intensify pollution further, the Soviets used Vosrazhdenya island in the middle of the Aral as their primary chemical weapons testing facility.

As a consequence of these pressures, the decline of the Aral Sea has been precipitous and terminal. Wildlife in the region has been eliminated and the local fishing industry destroyed. In addition, the Sea has declined in area from 66,900 square kilometres in 1960, to 33,800 square kilometres in 1991. As a consequence, the main fishing port on the Kazakstani side, Aralsk, has been stranded 100 kilometres from the new waterline. Wind erosion from the newly exposed seabed creates regular dust storms which deposit salty and chemically polluted sand over huge areas of land, salinizing the soil and rendering it agriculturally useless. Reduction in the Sea's size has also caused localized climate change, resulting in hotter summers and colder winters, which have shortened the growing season for crops. As can be imagined, the effect on the health of the local population has also been catastrophic. Infant mortality in the region has been reported at greater than 100 per thousand and diseases of the stomach, intestines and liver are all endemic.[2] Mothers have been warned not to breast-feed their infants because of the high concentrations of toxins likely to be found in their own systems and the rate of oesophageal cancer in the area was seven times the Soviet average in the late 1980s.[3]

Kazakstan's second major environmental disaster area results from the Soviet nuclear weapons programme of the Cold War. Moscow's most important nuclear test facility was situated at the 'Polygon' in north-eastern Kazakstan. During the Cold War period, the Soviet military tested nuclear weapons at the facility almost without restriction. Since 1948, the test site has witnessed around 500 nuclear explosions, including 160 atmospheric detonations. After the Soviet Union's ratification of the Partial Test Ban treaty in 1963, above-ground tests were outlawed, though poor safeguards resulted in roughly one out of every three of the subsequent underground tests venting at surface level.[4]

The effects of such an unrestrained nuclear test programme on the surrounding environment and population have been catastrophic. The implications for civilian contamination have also been intensified by the close proximity of Semipalatinsk, a city of around 340,000 people situated only around 150 kilometres from the test site. Though precise data on the extent of radioactive pollution and its effect on human health are unavailable (owing largely to the continuing silence of the Kazakstani governing elites), what information does exist suggests a human tragedy whose legacy will stretch for generations.

Data collected by the anti-nuclear organization 'Nevada-Semipalatinsk' in the late 1980s, and based on local medical studies, presented a sobering picture of the health problems prevalent in the regions adjoining the Polygon. It suggested that, among other things, the number of people suffering from malignant tumours was approximately 3.5 times higher in districts next to the Polygon than in other areas of the country, and that leukaemia rates among local children were significantly higher than the national average. Similarly, local statistics for the number of stillborn children, or those born with birth-defects, point to abnormally high rates in Semipalatinsk *Oblast*[5]. Local doctors also reported that in the ten days immediately following an underground test on 16 October 1987, the admission rate to district hospitals was 20 per cent higher than in the ten days prior to the test.[6] Bhavna Dave, writing in 1997, suggests that around 1.5 to 2 million people were affected by radiation in the region. She notes that the immune systems of 60–70 per cent of local inhabitants appear to have been undermined and that around 95 per cent of children in the area are thought to suffer from anaemia. Both these factors have led to the existence of a localized specific immune deficiency known as 'Semipalatinsk AIDS'. Indeed, Dave observes that some research suggests that the mortality rate in Semipalatinsk *Oblast* is 40 per cent higher than in the rest of the Republic.[7]

While the desiccation of the Aral Sea and the radioactive contamination around Semipalatinsk are perhaps the two most significant environmental problems Kazakstan faces, they are not the only ones. Extensive nuclear weapons tests were also carried out in other regions of the Republic, for example, at Kyzyl-Orda and Taldy-Kurgan. The main Chinese nuclear test site also lies only a few hundred kilometres east of the Kazakstani border, at Lop Nor in the Taklamakan desert. Moreover, the Republic suffers from several other environmental problem areas which will only be mentioned briefly here. They include very high levels of air and water pollution. In north-eastern and central areas of the Republic, for example, around one tonne of pollutants were released into the atmosphere per person in 1988.[8] Polluted run-off from industry and agriculture affects many of Kazakstan's rivers, rendering drinking water in some regions unsafe. Additionally, both Lake Balkhash and the Caspian Sea suffer from high levels of industrial pollution. In Lake Balkhash, the construction of the Qapchagay reservoir along its tributary river the Ili, and discharge from ore processing factories along its shoreline, have resulted in increased salinity and damage to local floral and fauna. Similarly, the Caspian suffers from pollution from both the Volga and Ural rivers, which has resulted in damage to the area's fishing industry and a decline in caviar stocks.[9]

Awareness of Kazakstan's unenviable environmental situation began to grow in the mid-1980s, in conjunction with a similar process in the other areas of the Soviet Union. The catalyst which stimulated the development of this awareness was the Chernobyl disaster of 1986. In the new atmosphere created by President Gorbachev's Glasnost policies, Soviet citizens found themselves in a position increasingly to question their government's environmental record.

Before Chernobyl, it is interesting to observe how little the environment influenced Soviet policy as an issue, despite the fact that the Soviet Union had a tradition of limited, localized, environmental activism dating back to the 1960s.[10] M. Goldman, in a 1992 study of Soviet attitudes towards the environment, notes that,

> Before 1985 and Gorbachev's election as General Secretary, pollution was a relatively minor concern. Soviet officials and academics insisted that for the most part, pollution was a capitalist, not a socialist, problem. It was the inevitable result of private corporations pushing off their costs into the public sector. Since the Soviet Union had no private corporations, by definition there could be no pollution. The public sector in the Soviet Union would not knowingly pollute itself and, after all, everything was part of the public sector.[11]

Indeed, on the surface, the Soviet Union had possessed stringent environmental protection laws since the time of Lenin. An example includes laws on factory emissions which were stricter than those imposed by the United States Clean Air Act of 1970 by a factor of four.[12] In practice, however, such laws were routinely ignored and flouted, and environmental protection was routinely sacrificed in the name of economic development. It is salient to note, for example, that Soviet scientists predicted the future desiccation of the Aral Sea (and even some of the more extreme consequences of such a situation) as early as the 1960s and that measured 'cost/benefit' decisions were made with regard to river flow into the Sea. Ultimately, it was argued that the rapid development of a cotton monoculture in the Aral Sea basin justified the sacrifice of the Aral Sea itself and all the problems this would entail.[13]

Kazakstan particularly had seen a huge programme of environmentally careless industrialization and 'modernization'. In agriculture, for example, Khrushchev initiated the so-called 'Virgin Lands' programme, which resulted in vast tracts of 'idle' land in northern Kazakstan, southern Russia and Siberia being put under the plough. In practice, the inept manner with which Virgin Lands was implemented, and the intensive nature of the farming practices used, meant that the consequences for the Kazakstani environment were extremely negative. The 'idle' land earmarked for cultivation was often actually highly productive pastureland, which supported Kazakstan's traditional livestock-breeding economy. In addition, reckless farming techniques led to the rapid decline in the productivity of much of the land concerned. By 1961, for example, the harvest was 30 per cent down on that of 1958. The harvest of 1962 was worse again.[14]

At this time, of course, popular attitudes towards the environment were often shaped by the propaganda of the Soviet regime, and as such frequently appeared unaffected by environmental concerns. Akiner observes that the popular image of Kazakstan at the time of the Virgin Lands programme was as 'A Storehouse of Natural Riches . . . A Republic of Major Industries . . . of Collective Farms and State Farms . . . A Land with a Great Future'.[15] Similarly, Soviet propaganda had earlier described how the local population in Uzbekistan had happily flocked to give their labour free during the construction of irrigation canals.[16] However, while an estimated one million Soviet citizens did respond to Khrushchev's appeal for Virgin Lands workers, and participated wholeheartedly in the implementation of Kazakstan's agricultural transformation, these 'enthusiasts' were made up almost exclusively of Slavs from the western regions of the Soviet Union. In contrast, there is a great deal of evidence which suggests that many ethnic Kazaks were extremely disconcerted by the potential impact of Moscow's initiatives on the economy and environment of northern Kazakstan. W. Kolarz, for example, observed that resistance to the programme among the Kazak elements of the Communist Party of Kazakstan was so pronounced that a fundamental reshuffle of personnel at the highest level was ordered by Moscow. Indeed, Zhumabay Shayakhmetov, the first ever Kazak First Party Secretary of the CPKZ, was dismissed in February 1954, to be replaced by P. K. Ponomarenko, a Russian, after disagreements over the implementation of Soviet policy in Kazakstan.[17] Similarly, the influential Moscow journal *Kommunist* virulently attacked the persistence of nationalism among Kazak intellectuals in an article in 1959, largely in response to their criticisms of Moscow's Virgin Lands scheme.[18]

Despite these local rumblings of discontent, however, Moscow's economic policies remained largely uninfluenced by environmental concerns. In addition, much of the nationalist-orientated opposition to Moscow's various development projects in Kazakstan was dissipated after the appointment of D. A. Kunaev, a Kazak, to the position of First Party Secretary of the CPKZ. Kunaev helped to restore a degree of equality into Kazakstan's relationship with the Soviet centre, and established ethnic Kazak dominance over certain sectors of the Republic's economy. In this changed ethno-political situation, environmental concerns lost some of their nationalist potency, and receded into the political background. As a consequence, Soviet thinking until Chernobyl, at both national and republican levels, continued to see the environment as a tool to be used in the economic growth of the socialist state, however unsustainable the benefit gained would be in the long term.

Public attitudes towards environmental problems in the Soviet Union shifted considerably in the wake of the Chernobyl disaster, to the extent that, by 1990, it has been estimated that well in excess of 300 pro-environmental and anti-nuclear groups were in existence in the USSR.[19] The devastation caused by Chernobyl, and the subsequent burst of misinformation about the disaster spread by the Soviet regime, shook many people's faith in the credibility and competence of their government, at least as far as pronouncements on environmental issues were concerned. A Kazak environmentalist I interviewed during the summer of 1996 explained his own, and his family's, reaction to the Chernobyl disaster as more information began to seep out in the late 1980s. He recalled,

> I was a 'Pioneer'[20] at the time of Chernobyl, and a good Communist. When the disaster happened, the government said, 'Volunteer to help the people of Chernobyl. Together we can all help'. I volunteered and went to Chernobyl with many of my friends. We were all involved in the clean-up operation in some of the outer zones . . . When we returned, it became clearer what had happened at Chernobyl. My father was so angry at the time. He couldn't believe the government had sent its own children into such a dangerous area . . . To this day my mother doesn't know that I went to Chernobyl, and there isn't a day that goes by when I don't think about it.[21]

Indeed, the new climate of openness engendered by Glasnost allowed Soviet citizens to see exactly the extent to which their government had lied to them. Increasingly, it was possible to disseminate and discuss information which in the past would have been restricted. In this way general awareness of the problems the Soviet Union was facing in the ecological sphere rose significantly. Moreover, increasing political freedoms in the USSR created the conditions by which those who wanted to protest could, albeit in an initially low-key manner.

Environmental organizations began to spring up throughout the Soviet Union in rapidly increasing numbers. They focused almost exclusively on local issues, such as concerns about a nuclear power station, or worries over air pollution from factories. As a result, the Soviet environmental groups of this period formed a disparate, *ad hoc* collection of small 'Not in My Back Yard' (NIMBY) organizations. In general these movements were led by members of the cultural and scientific intelligentsia. Additionally, it is interesting to compare Soviet environmentalism with the wave of environmental activism which was also sweeping many western states at this time. In contrast to the western ecological movement, there was very little focus on global issues among Soviet groups and very little co-ordination between organizations. J. DeBardeleben observes that most groups were small, with a core of two to five individuals and a broader membership circle of twenty to thirty.

She notes that, in the late 1980s, it was typical for such an organization to be able to rally several hundred people in protest against a particular local problem.[22]

Despite these limitations however, environmentalism in the Soviet Union increasingly came to represent a means by which Soviet citizens could participate in political activities outside of the officially sanctioned Party framework. For the first time, the limited democratization in Soviet society had introduced ideas of political accountability and responsiveness which had not been present before. In the republics, and certainly in Kazakstan, the responsibility for many local environmental problems was directed to the Soviet centre, Moscow. For many Kazak intellectuals, for example, the ecological devastation at Semipalatinsk was indicative of Moscow's colonial and exploitative attitude towards their region.

Indeed, the local nature of the environmental protests, and the tendency for the blame to be laid at the feet of a distant administration in the Soviet capital helps to explain why environmental protests were tolerated in a way in which, for example, overt nationalism was not. J. Dawson argues that the growth of popular environmental protest in the Soviet Union was tolerated by elements of the Soviet administration because it was not directly critical of the foundations of the Soviet state or governmental system. She notes that while Moscow generally viewed the new movements with trepidation, regional and republican administrative structures often saw environmentalism as a safe, apolitical outlet for popular frustrations and public participation. In addition, because it was Moscow that held responsibility for energy production and distribution, and because often it was precisely this issue which stimulated the protests, local officials (who themselves were becoming increasingly dependent on public goodwill for their continued tenure in office) were inclined to look upon such actions relatively benignly.[23]

The official tolerance enjoyed by environmental movements in the Soviet Union at this time stimulated their growth in importance as foci for expressions of popular discontent over a whole range of issues. Perhaps most significantly, environmentalism became an important element of many of the republican nationalist movements which were beginning to emerge. Environmental activists often saw their own political roles expand. M. Goldman observes that the initial willingness of these activists to 'stick their heads above the parapet' greatly increased their popular credibility. He notes that often such activists were viewed as 'critics who could be trusted to speak out', who, after Glasnost's relaxation of restrictions on organizations, became the natural leaders of the new political groupings.[24] In many cases, environmental protest in the Soviet Union was also closely linked to other issue areas and often to nationalism. Indeed, many western analysts have suggested that the growth of environmentalism in the USSR in the late 1980s was little more than a *surrogate* for the growth of nationalism. These trends will be further explored below, using the Kazakstani experience as a case study.

Kazakstan had shared many of the trends visible throughout the Soviet Union during the Glasnost period, though, as with many of the other Central Asian republics, the democratization process had proceeded somewhat more slowly than in the more western republics. Increased freedom of information had stimulated a general (though still relatively limited) growth in public environmental consciousness. The extent of the Republic's ecological problems were, as a result, beginning to come to the attention of certain elements of society. In addition, the legacy of Chernobyl, and the activities of environmental groups in other areas of the USSR, had illustrated to many people just how little the government could be trusted as far as its environmental assurances were concerned. In Kazakstan, the environmental protests of the late 1980s were initially sparked by a general outpouring of public disgust towards the Soviet Union's nuclear weapons tests at the 'Polygon'. The Kazakstani environmental movement which would emerge as a result would play a critical part in the development of the post-independence Kazakstani political system.

The 'crisis point' which finally launched the Kazakstani environmental movement was reached between 12 and 17 February 1989. Over this period, the Soviet military carried out a series of nuclear tests at Semipalatinsk, with the result that radioactive fallout levels in the region increased dramatically and noticeably.[25] In response to these actions, the renowned Kazak poet and author Oljas Suleimenov made an open appeal on Kazakstani television on 26 February for the cessation of nuclear testing in Kazakstan. Suleimenov's commitment to civil rights issues was well know and respected throughout Kazakstan, by all ethnic groups. While he had acquired a reputation as something of a Kazak nationalist (his most important work, *Az i ya*,[26] stressed the Turkic contribution to Slavic culture and attracted a barrage of criticism from Moscow in 1975), S. Akiner emphasizes that he was 'a Kazakh who spoke and wrote in Russian and was in many ways more at home in a Russian environment than a Kazak one'.[27] As such Suleimenov retained credibility among both Kazaks and Russians, a factor which lent an impact to his protest that other figures may not have had. Over the next couple of days, meetings were held between concerned Kazakstani intellectuals, culminating in the creation of an environmental pressure group, named after the most significant US and Soviet nuclear test areas: 'Nevada-Semipalatinsk'.

The public response to the establishment of the group was immediate and overwhelming. Within the next few months the movement had collected a petition of over a million signatures calling for an end to nuclear testing as well as numerous charitable donations.[28] Thousands of letters and telegrams expressing support for the aims of Nevada-Semipalatinsk were also received, and, by May, contacts had been established with western anti-nuclear movements. Against this background of protest, however, the government resumed nuclear testing in July, an action which served to garner further support for Kazakstan's anti-nuclear movement. The atmosphere of the time was described to me by a Kazakstani environmental activist who participated in many of Nevada-Semipalatinsk's protests. He remembered that, 'at that time it was very easy to gather people for round-table discussions and conferences about ecology in Kazakstan. Everyone was interested and many people were willing to stand up and say "no" to the Polygon.'[29]

Throughout 1989, Nevada-Semipalatinsk organized a series of demonstrations and other actions (including, for example, a protest at the very border of the test site itself) and succeeded in preventing eleven of the eighteen nuclear tests planned for that year. Indeed, the nuclear test at Semipalatinsk on 19 October 1989 proved to be the last conducted by the Soviet Union. In October 1990, the USSR became the first nuclear state to announce a moratorium on nuclear testing. Soon afterwards, in December, the collapse of the Soviet Union and the advent of independence for Kazakstan signalled the death knell of the Polygon. On 29 August 1991, the Kazakstani President, Nursultan Nazarbaev, signed a decree, closing the test facilities at Semipalatinsk for good.[30]

The reasons for the overwhelming successes of Nevada-Semipalatinsk in 1989 stem from several areas. First, it should not be forgotten that the scale of the suffering experienced by the inhabitants of the Semipalatinsk region as a result of nuclear testing was considerable, and, with the help of Nevada-Semipalatinsk, had attained considerable public sympathy. In one 10,000 signature petition from 'the inhabitants of Egindybulak *Raion*, Karaganda *Oblast*', for example, local people claimed that 'it is impossible to convey the horror and sorrow of parents who, instead of the impatiently awaited healthy child, see a two-headed, arm-less, disfigured creature. Most of our people do not live beyond the ages of fifty or sixty. Anxiety about future children, heirs, tortures us.'[31] In the face of such heartfelt declarations, a certain degree of public support could be assured. After all, the inhabitants of the Semipalatinsk and Karaganda *Oblasti* were fellow Soviet citizens, and, if they could be treated in such a way, then what was to stop the government from doing so again to other sections of the population?[32] However, as other human tragedies such as Chernobyl have illustrated, the justness of a cause, or the degree of suffering inflicted on the victims of a disaster, does not necessarily guarantee the political success of a protest movement.

Nevada-Semipalatinsk, and indeed environmentalism in Kazakstan more widely, also managed to tap into other trends in Kazakstani society whose importance was beginning to emerge. Of these, it is the strong linkage Kazakstani environmentalism has with *Kazak* nationalism and national identity that is perhaps the most significant. Environmentalism and nationalism in the Soviet context make comfortable bedfellows. The environmental damage inflicted on the Kazakstani environment, at least in the popular imagination, was damage inflicted *on* Kazakstan, *by* Moscow. As such, it could easily be portrayed as a violation of *national* territory, or even of national culture, by an alien foreign power. This tendency to view Moscow's role in the degradation of the Kazakstani environment in this light was also reinforced by the history of the Kazaks' relationship with the Soviet (and earlier Russian) state, which included near genocidal episodes such as the collectivization and settlement of the Kazak nomads.[33] Among Kazak nationalist groups today, a strong resentment of Soviet environmental damage remains a central ideological plank. When I interviewed A. A. Amirbekov, head of the South Kazakstan *Oblast* branch of the nationalist party Azat in 1996, for example, he complained that: 'We fought to close the nuclear test site at Semipalatinsk . . . During Soviet times, there were communist rules, Moscow rules. The Polygon belonged to Moscow. It was total discrimination against all nations by Moscow.'[34]

In addition, a close relationship with the environment plays an important role in modern perceptions of what makes up Kazak national identity. Though many Kazaks (and certainly most Kazak politicians and academics) are today city-dwellers, Kazak 'culture' retains strong links with the Kazak's nomadic past. Indeed, before their forced collectivization and settlement in the 1920s and 1930s, the vast majority of Kazaks were practising nomads. Settled Kazak culture is still a relatively modern phenomenon. While the latter part of the Soviet period saw a destruction of many traditional elements of Kazak society, and an extremely strong assimilation of Soviet identity among the Kazaks, 'nomadism' remained one of the defining criteria of 'what it meant to be a Kazak'.

Central to nomadic existence is a symbiotic relationship with, and dependence on, the surrounding environment. As such, a crucial element of the Kazak's 'nomadic' identity is seen to be consideration for the environment. In practice of course, and perhaps excluding the nationalist-orientated discontent of the Virgin Lands period, Kazak officials in Soviet times often showed no more respect for the environment than their Russian counterparts. What is important in the context of environmental activism, however, is that a perception exists among Kazaks today that environmentalism is part of their 'national character'. This factor makes the linkage between ecological activism and Kazak nationalism all the easier. It is interesting to note, for example, that the design of the new Kazakstani flag incorporates strong environmental symbolism. Akiner describes it as including

> a band of Kazakh ornamental motifs and a representation of the Steppe eagle, favoured hunting companion of the Kazaks of old; the main elements of the coat-of-arms are the winged horses of Kazak myth and the sacred smoke-hole of the yurt. The national colours of blue and gold, representing the sky and the sun, have a universal significance, but also a symbolic link to the ancient Kazakh cult of the Sky God.[35]

Despite the strongly Kazak nationalist elements in the Kazakstani environmental movement during this period, however, Nevada-Semipalatinsk always stressed the multi-ethnic nature of its aims. In reality, the organization (and its sister group, 'Aral-Balkhash', led by Mustapher Shakhanov) was an extremely broad church, uniting many different political creeds and colours. Many Nevada-Semipalatinsk activists were Russians, for example. In addition, it was

Kazakstan's Slavic population which had suffered the brunt of Soviet nuclear testing, Semipalatinsk being an overwhelmingly Russian-dominated city. Indeed, the sheer variety of political viewpoints represented within Nevada-Semipalatinsk was ultimately to lead to splits in the movement. From within the Nevada-Semipalatinsk movement of 1989, there emerged the foundations for many elements of contemporary Kazakstani political parties. As such, it can be argued that environmentalism, and the Nevada-Semipalatinsk movement specifically, created the foundations for the development of a 'civil society' in the new Kazakstani state.

Cracks began to emerge in the Nevada-Semipalatinsk coalition in 1990. While the organization continued to press for the formal closure of the Polygon test site, the test moratorium declared by the Soviet leadership had succeeded in defusing much of the passion initially created by the protests. Besides, other issue areas were beginning to be prioritized by certain activists within the movement and divisions between campaigners were increasingly vocalized. Many Russians attacked the perceived Kazak orientation of certain areas of Nevada-Semipalatinsk's activities. They argued that the protests against nuclear tests had much more to do with the problems of Russians in the region than with Kazak nationalism, and that the nationalist orientation of Nevada-Semipalatinsk ensured that it could not adequately represent the real victims of the Polygon.[36] In contrast, Kazak nationalists voiced their own complaints, criticizing Oljas Suleimenov's practice of writing and speaking in the Russian rather than the Kazak language.

Almost inevitably, the divisions within Nevada-Semipalatinsk became formalized, as various activists split from the main body of the organization to create their own splinter groups. It was these new groups which often formed the basis for the creation of Kazakstan's new generation of political parties. Examples of Kazakstani political groupings whose origins lie in the so-called 'Green Front' of Nevada-Semipalatinsk in 1989 include: Azat (a Kazak nationalist party founded in June 1990), the Republican Party of Kazakstan (itself a moderate splinter group from Azat), and Attan (whose leader Amantai Asilbekov combines environmental, Kazak nationalist and Islamic rhetoric).[37] As well as these identifiable Nevada-Semipalatinsk splinter groups, numerous figures active in the Kazakstani opposition movement can claim involvement in the 'Green Front' of 1989. These include members of pro-Russian groups such as 'Lad', as well as activists in the Socialist Party of Kazakstan and political coalitions such as 'Azamat'.[38]

Nevada-Semipalatinsk itself, while weakened by these divisions and defections, entered the new world of independent Kazakstani politics as the most significant and powerful political actor outside of the Republic's established power structures. Oljas Suleimenov still commanded enormous respect from the Kazakstani population and was regularly mooted as the only genuine contender for Kazakstan's presidency aside from the incumbent Nursultan Nazarbaev. In order to help preserve the purely environmental focus of

Nevada-Semipalatinsk, Suleimenov went on to found a more overtly political wing of the movement: 'People's Congress'. People's Congress retained links with Nevada-Semipalatinsk but campaigned on a broader, centrist political agenda and, by 1992, had become a critical player in the Kazakstani political process. Conversely, Nevada-Semipalatinsk as an environmental group attempted strenuously to distance itself from the twists and turns of everyday politics. In an interview I conducted with M. A. Abishev, vice-president of Nevada-Semipalatinsk in 1996, for example, it was stressed that the group was 'a non-political organization. Our main aim was to close the Polygon. We think that our activity had an influence on the closure of all nuclear test sites. Now our present activity is to help people who have been irradiated by nuclear tests.'[39]

Indeed, Nevada-Semipalatinsk's anti-nuclear campaign of 1989 did not simply have political consequences. In a very real way, it stimulated the environmental debate in Kazakstan and brought the whole gamut of the Republic's ecological problems to public attention. One of the most noticeable aspect of the Kazakstani 'Green Front' was that it successfully decentralized much of its activity, and, as a result, established local branches (and disseminated information) in some of the most remote regions of the Republic. In fact, it could be argued that because of the immense size of Kazakstan, and the relative sparseness of its population, such an approach was the only way of protesting effectively. Today, these local branches have stimulated the growth of issue specific, local environmental pressure groups who focus and campaign on regional problems. Examples of this type of organization are numerous, but include active groups as diverse as 'Children of the Aral', 'Ecochance', 'Ecoinform', 'Ecolog Club', 'The Foundation for Support of Ecological Education', 'Green Future of Rudni Altai', 'Green Women', 'Greenspace' and the 'Kokjiek Society for Aral Sea Region Problems'. Often these organizations are small, with few resources, but many, such as 'Kokjiek' and 'Ecoinform', have close contacts with western and international organizations such as the United Nations Development Programme (UNDP) and the American Peace Corps. Some, such as 'Children of the Aral' are closely connected to Kazakstan's Ministry of Health Protection.[40]

Additionally, the close association of many of Kazakstan's most respected literary and cultural figures with the Nevada-Semipalatinsk campaigns, as well as the depth of Kazakstan's environmental problems, has lent environmental politics in Kazakstan an air of the moral high ground. Espousing elements of environmentalism in a political party's agenda has often acted as a means of legitimizing its political credentials. As groups such as Azat have shown, environmentalism can form an effective badge of respectability, often serving to 'take the edge off' the impact of any more extreme policies any party may have. In my own experience, almost every opposition political party or organization made some sort of (often somewhat confused) environmental claim, usually early on in any interview. The pro-Kazak group 'Jeltoqsan', for example, made a great play of arguing that Kazakstan's environmental problems were the fault of Moscow, and, by extension, the fault of the Russians. Jeltoqsan, it was claimed, was pro-environment and therefore pro-Kazak.[41] Conversely, R. I. Ivanitsky, vice-president of the Slavic movement Lad, used the environment as a reason to unite Kazakstan's ethnic groups. He opined that 'the ecological problems facing the population are the common problems of everybody here, of all the people of Kazakstan where the "Russians" are only part of the entire multi-national people of Kazakstan'.[42]

In the years since Kazakstan achieved independence, however, the actual influence and impact of the environmental movement has declined steadily. Much of the political threat posed by People's Congress has been neutralized by the effective political exiling of its popular leader, Oljas Suleimenov. Amid various allegations of financial misdealing by Nevada-Semipalatinsk, Suleimenov was offered the governmental position of Ambassador to Italy in June 1995 (conveniently coinciding with Kazakstan's parliamentary elections and several important constitutional referenda which would serve to strengthen the position of the incumbent president). Given the threat of more explicit governmental action over the allegations, he had little option other than to accept the posting.[43] This development succeeded in effectively isolating Suleimenov from opposition activities by physically removing him from Kazakstan, while also placing him in an official governmental position from which anti-government political activity was virtually impossible. As such, People's Congress, and Nevada-Semipalatinsk more broadly, found itself without its figurehead and leader, and effectively decapitated. While the departure of Suleimenov did not affect the actual ecological work of many of Nevada-Semipalatinsk's local branches, or indeed of Aral-Balkhash, it did eliminate much of its political influence.

The reasons for a broader decline in political environmentalism in Kazakstan are twofold however. The first problem derives from weaknesses in many of the organizations themselves. The second results from political obstacles imposed by the established power structure. Both these problems are shared by most other opposition parties and movements in the Republic today, from all political backgrounds.

Most non-establishment environmental groups and political parties in Kazakstan suffer from structural weaknesses and financial problems. In many cases parties and organizations are run solely by a small hard-core of activists (sometimes by only one or two people), with extremely limited resources and infrastructure. In this sense, very little has changed since the initial establishment of the Soviet environmental movement in the mid-1980s. The Civic Peace Association, for example, which has campaigned on raising national and international awareness of Kazakstani environmental problems has been forced to operate for part of the time as a tourist firm in order to help finance its activities.[44] Additionally, and partly as a result of their lack of funding, most organizations have minimal access to the means by which party literature can be disseminated such as printing presses or air and radio time. Complaints about such a lack of resources were common among the majority of organizations to whom the author spoke during 1996, from all sides of the political spectrum. It was also rare to find a group with a clear national platform or even with reliable links to party activists in other regions of the country. In my own experience, regional activists would often espouse different policies from those promoted in Almaty and would sometimes be unaware of some of the main tenets of their party's policies.[45]

Many of these problems were, and still are, compounded by the actions of the government, which, in some cases at least, has directly obstructed the activity of organizations which are perceived as forming the political opposition. Because of the prominent role environmentalism played in the late 1980s and early 1990s, the ecological movement has found itself a prime target of such governmental pressure. Many parties and organizations have faced considerable constitutional barriers before they can even operate legally and several activists from different organizations told stories of more direct state pressure such as police intimidation. Mels Yeleusizov, president of the environmental party 'Tabighat' (meaning 'Ecology' in Kazak), described his position to me. He observed:

> There is pressure, constant pressure. Dirty Tricks. For example, my telephone has been switched off. Several times they tried to evict me from my office, and during the referendums[46] I know that there was a list of people who were forbidden from appearing or speaking publicly. Our mass media were under great pressure: they didn't even write bad things about me. Even my friends asked me 'where did you get lost?' But I was here . . . There is a political committee in our party. The members are intellectuals. When we started they [the government] started to put pressure on me. One time I was beaten. Several times they tried to buy me, but when they understood that they couldn't stop me, they started to put pressure on members of our political committee, my friends. Now I'm practically alone. I had many supporters, but most of them have disappeared because of the state pressure.[47]

The government has also attempted to clamp down on political opposition through statutory means. By law, all parties and organizations in Kazakstan must be formally registered before they can engage in political activities or contest elections. The basic criteria for registration is that the organization commits itself to the Constitution of the Republic and to 'democratic values'. Registration itself is a long and bureaucratic process and in many cases parties and organizations have been declared illegal because some of their policies have been deemed 'unconstitutional'. Members of a Russian organization complained in conversation with me, for example, that their registration had been refused on the grounds that some of their aims 'opposed the sovereignty of the Republic of Kazakstan', despite the fact that all they had done was to call for closer integration with the Russian Federation.[48] If an application is refused, the organization concerned must reapply, with the new process taking many months or even years. Certainly, during the 1995 parliamentary elections, registration was used with great effect to eliminate much of the opposition (Tabighat, for example, remained, and remains, unregistered). Additionally, a registration fee of Tg30,000 ($500, equal to the average annual salary at the time) was imposed on all prospective candidates.[49] This tactic succeeded in preventing most independent candidates from even attempting to stand for election.

So, while the environmental movement in Kazakstan functioned as a 'proving ground' for many of the most prominent members of the Republic's opposition, it has also suffered as a result of its past. Coupled with this problem, the Kazakstani government has, in recent years, increasingly distanced itself from environmental issues generally, with the result that ecologically orientated pressure groups have become even more excluded from the mainstream political process. The administration has shown itself to be decidedly reluctant to commit funds to, or even attempt to tackle, some of Kazakstan's more intransigent ecological difficulties. In truth, the government of Kazakstan has found that its political priorities lie in areas such as economic development rather than environmental protection and regeneration. More often than not, these two issue areas conflict and, when this has happened (as with the development of the Tengiz oil-fields in north-western Kazakstan), it has generally been the environment which has lost out.

Additionally, some problems, such as that of the Aral Sea, are so enormous that the government has simply not been willing (or able) unilaterally to commit the necessary funds or make the necessary sacrifices to deal with them. Likewise, the administration clamped down on protests against Chinese nuclear tests at Lop Nor for fear of damaging Chinese/Kazakstani relations. Amantai Asilbekov, the leader of the organization Attan, for example, had recently been released from temporary house-arrest when I interviewed him in July 1996. His confinement had lasted the duration of a visit by the Chinese Premier to the (then) Kazakstani capital,[50] and had been instigated to prevent any embarrassment Asilbekov may have caused by protesting against Chinese bomb tests.[51]

Similarly, practical (if not emotional) popular support for many environmental groups has declined, especially in those areas less directly affected by environmental problems. Since independence, many Kazakstani citizens have found that their priorities have changed. The difficulties experienced by all sectors of the Republic's economy since independence have led to a dramatic decline in the standard of living for the majority of Kazakstan's population. Many Kazakstanis make barely enough money from their wages to cover their food bills, let alone other expenses such as rent or utilities.[52] The provision of utilities has also declined in quality since the collapse of the Soviet Union. Even in the relatively prosperous Almaty suburbs in 1996, for example, people often lacked gas or hot water for maybe three days a week. These difficulties multiplied in the more remote cities such as Karaganda, where one interviewee complained to me that she had been without gas for over seven months.[53]

In conditions such as these, many Kazakstani citizens have been reluctant to campaign for environmental protection measures, particularly if these may directly affect the local economy or employment prospects, for example through shutting down a polluting factory. In Petropavlovsk, a representative of the local branch of Lad opined to me that the government should actively strive to reopen the city's nuclear missile factories (which had closed after the collapse of the USSR) in order to create employment and stimulate the local economy.[54]

Likewise, common responses among people in Kazakstan to questions about the political situation included comments along the lines 'we have too many problems getting food on the table at the moment to worry about politics', or 'Nazarbaev may have his faults, but we know what they are. Why should we risk putting someone else in power after all the changes that happened last time we did it?'[55] Similar results emerged in a survey conducted by the Academy of Sciences in Almaty. When respondents were asked what in their opinion were the most important attributes for an independent Kazakstan, the highest responses included 'stability and a normal level of life', 'family wealth' and 'peace and stability in the Republic'. 'Ecological safety of society' featured considerably lower down on the list.[56] These trends were also recognized by Kazakstani environmental groups themselves. M. I. Abishev, of Nevada-Semipalatinsk observed to me that, 'When we started our movement, we had millions of followers in Kazakstan, and many in the international community. Now because we have implemented our main aim, and because of today's economic crisis, it is less. People think about day-to-day life and they can't get involved in politics.'[57]

Environmentalism in Kazakstan remains important however, not least because the Soviet legacy of environmental damage survives in a very real way. In the aforementioned survey, environmental issues may have been perceived as less important than stability and financial security, but, given most people's current economic situation, this result is hardly surprising. What is perhaps more illuminating is a different part of the same survey which focused specifically on ecological issues. In this section, 99 per cent of all respondents declared that ecological problems troubled them. Even more interesting was the claim made by some 50 per cent of respondents that unfavourable ecological factors had influenced both their and their relatives' health.[58]

Moreover, whatever the Kazakstani government's policies have been in reality, it has remained necessary for it to continue vocally to support environmental issues areas, at least on the surface. It has, for example, established a specialist Ministry for the Environment as well as numerous environmental protection agencies. Kazakstan has also participated in a variety of international conferences and meetings over the Aral Sea crisis, in an attempt to co-ordinate some form of regional response to the problem. Though some cynics suggest that international grants to tackle the Republic's environmental problems often appear to have been 'siphoned off' by local officials, and that the Aral Sea conferences have resulted in very little of concrete benefit to the problem, the fact that the government has felt the need to create such a high-profile environmental stance is a reminder of the importance of ecology in the popular Kazakstani psyche.[59]

It is also important to remember that the environmental movement in Kazakstan has a track record of success that many western organizations would dearly love to emulate. It did play a significant role in the cessation of nuclear testing in the Soviet Union, and has succeeded in bringing ecological issues to the national political agenda. As a result, a residual feeling appears to remain among the Kazakstani population that the environment is one area in which ordinary people can influence national policy. In practice this attitude has translated into real successes for many of the smaller, localized groups. In Ust-Kamenogorsk in East-Kazakstan *Oblast*, for example, the officially registered Green Party of East Kazakstan has, since 1992, managed to raise sufficient funds to install a scrubber in the chimney of a local asphalt plant, initiated a tree planting programme, and established a thriving environmental information centre in the city.[60] American Peace Corps volunteers involved with the project observed to me that the East Kazakstan Green Party had genuinely served as a catalyst for grassroots environmental activism in the *Oblast*.[61]

Likewise, the environmental movement nationally has formed the basis for Kazakstan's embryonic non-governmental political sector, and, as such, has had a strong democratizing influence. Opposition circles in Kazakstan retain a marked environmental bias, and ecological worries feature high on most political parties' agendas in some form or another. The years since 1996 have seen new opposition coalitions begin to emerge, many of them, such as 'Azamat', consist of figures prominent in the 'Green Front' of 1989. Though not structured wholly around environmental politics, these new groupings often share many of the aims and principles of Nevada-Semipalatinsk in its heyday, and have begun to play an increasing role in Kazakstani politics.

The fate of the Kazakstani environment itself remains uncertain, however. Despite a popular and political sensitivity to environmental issues in the Republic, the hard fact of the matter remains that often Kazakstan's ecological problems are so enormous and intransigent that the government simply does not have the political will or economic muscle to tackle them effectively. While the closure of the various Soviet 'Polygons' on Kazakstani soil have insured that no further damage will be done by nuclear testing, contamination from past explosions was so great that their effects on the Kazakstani environment and population will be felt for a long time to come. Likewise, Central Asia's rapidly increasing population suggests that demands for agricultural irrigation in the region are unlikely to recede in the future. Though investment in improving irrigation infrastructure (for example by lining canals, etc.) would help to lesson the impact of this demand, such a financial commitment seems improbable when resources for the adequate maintenance of existing facilities have not been allocated. Similarly, it is unlikely that Kazakstani factories will invest in anti-pollution measures when they cannot even afford to pay wages to their workers.

Nonetheless, political, demographic and economic pressures may ultimately force regional governments to face up to Central Asia's ecological difficulties. Sustainable development in the region in the long term is dependent on some of the more fundamental causes of today's environmental damage (for example management of water resources) being properly addressed. In addition, though the influence of the Kazakstani environmental movement may have declined in recent years, it has not disappeared. There remain within Kazakstani society significant pro-environmental leanings, which will help to ensure that the Republic's ecological problems will not easily disappear from the political agenda.

The environmental legacy left by the Soviet Union in Kazakstan is one not simply of ecological damage, but of political protest and activism as well. With Kazakstan showing ever increasing tendencies towards authoritarianism, this past legacy may be one of the best hopes the Republic has of preserving a democratic future and ensuring that the environmental mistakes of the past are not repeated.

Notes

This chapter is based primarily on research which I conducted in Central Asia during the summer of 1996 as part of the fieldwork for my PhD. While in Kazakstan, I worked closely with locally based NGOs (particularly the Civic Peace Association) and conducted interviews with a wide variety of politicians and political groupings. The Republic's environmental problems were approached candidly by almost all interviewees. The environment seems to be one area of the Kazakstani political debate where people feel free to comment and criticize freely. Among establishment politicians, for example, ecological difficulties can be blamed on the policies of Moscow, thereby lessening any political repercussions for the present regime. For ecological groups in the political opposition, interviews with a western academic represented another opportunity to 'get their message across'. I believe my close association with various Kazakstani environmentally orientated organizations encouraged interviewees from the political opposition to talk more openly about their experiences. In all cases, however, interviewees have only been quoted and/or named when they gave their express permission for me to do so.

1 J. Critchlow, 'Central Asia: How to Pick Up the Pieces?', in J. DeBardeleben and J. Hannigan (eds), *Environmental Security and Quality after Communism: Eastern Europe and the Soviet Successor States* (Oxford, 1995), 147.
2 Ibid. 148.
3 M. Feshbach and A. Friendly Jr, *Ecocide in the USSR: Health and Nature under Siege* (New York, 1992), xii.
4 B. Dave, 'Kazakhstan Staggers under Its Nuclear Burden', *Transition*, 17 November 1997, 12.

5 An *Oblast* is a Russian term for an administrative district. At the time of independence, Kazakstan had nineteen *Oblasti*, though several have been merged in recent years.
6 H. B. Paksoy (ed.), *Central Asia Reader: The Rediscovery of History* (London, 1994), 178.
7 Dave, 'Kazakhstan Staggers', 13, 72.
8 D. R. Smith, 'Kazakhstan', in P. R. Pryde (ed.), *Environmental Resources and Constraints in the Former Soviet Republics* (Oxford, 1995), 261–4.
9 Ibid. 265.
10 For further details, see J. DeBardeleben, 'The New Politics in the USSR: The Case of the Environment', in M. Stewart (ed.), *The Soviet Environment: Problems, Policies and Politics* (Cambridge, 1992), 64–87.
11 M. I. Goldman, 'Environmentalism and Nationalism: An Unlikely Twist in an Unlikely Direction', in Stewart (ed.), *The Soviet Environment*, 1.
12 Ibid. 3.
13 M. H. Glanyz, A. Z. Rubinstein and I. Zonn, 'Tragedy in the Aral Sea Basin: Looking Back to Plan Ahead', in H. Malik (ed.), *Central Asia: Its Strategic Importance and Future Prospects* (New York, 1994), 174–9.
14 M. McCauley, *Krushchev and the Development of Soviet Agriculture: The Virgin Land Programme, 1953–1964* (London, 1976), 142.
15 S. Akiner, *The Formation of Kazakh Identity: From Tribe to Nation-State* (London, 1995), 51.
16 R. A. Davis and A. J. Steiger, *Soviet Asia* (London, 1943), 80.
17 W. Kolarz, *Communism and Colonialism* (London, 1964), 42–5.
18 Ibid. 45–8. Interestingly, one of the Kazak intellectuals implicitly criticized by *Kommunist* was Oljas Suleimenov. See below for further details on Suleimenov and his role in contemporary Kazakstani politics.
19 DeBardeleben, 'The New Politics', 73.
20 The 'Pioneers' were the Soviet youth organization, in many ways equivalent to the Scouts.
21 Author's interview, Almaty, Kazakstan, 9 August 1996.
22 DeBardeleben, 'The New Politics', 73–4.
23 J. I. Dawson, 'Anti-Nuclear Activism in the USSR and Its Successor States: A Surrogate for Nationalism?', *Environmental Politics*, 4 (1995), 449–50.
24 Goldman, 'Environmentalism and Nationalism', 5.
25 Paksoy (ed.), *Central Asia Reader*, 179.
26 *Az i Ya* is a pun on the Russian pronoun *ya* ('I') and 'Asia'.
27 Akiner, *The Formation*, 56–7.
28 Paksoy (ed.) *Central Asia Reader*, 179.
29 Author's interview, Almaty, Kazakstan, 8 July 1996.
30 Dave, 'Kazakhstan Struggles', 12.
31 The Inhabitants of Egindybulak *Raion*, Karaganda *Oblast*, Kazak SSR, 'Letter to the Nevada-Semipalatinsk Movement', in Paksoy (ed.), *Central Asia Reader*, 182.
32 Author's interview, Almaty, Kazakstan, 10 August 1996.
33 For a detailed analysis of Kazak history and the history of the Kazak–Russian relationship, see M. B. Olcott, *The Kazakhs* (Stanford, CA, 1987).
34 Author's interview, Shymkent, Kazakstan, 5 September 1996.
35 Akiner, *The Formation*, 61.
36 Author's interview, Petropavlovsk, Kazakstan, 5 August 1996.
37 Assotsiatsya sotsiologov i politologov g.Almaty, *Politicheskiye Partii i Dvizheniya (II)* (Almaty, 1996), 29–30.
38 Author's interviews, various, Kazakstan, summer 1996.
39 Author's interview, Almaty, Kazakstan, 10 September 1996.

40 United Nations Development Programme, *Aral Sea Basin NGO Directory: Environmental and Health Protection Non-governmental Organisations in Kazakhstan, Kyrgytstan, Tajikistan, Turkmenistan, Uzbekistan* (Tashkent, 1996), 14–18, 22–25, 32.
41 Author's interview, Almaty, Kazakstan, 8 July 1996.
42 Author's interview, Almaty, Kazakstan, 11 September 1996.
43 B. Dave, 'A New Parliament Consolidates Its Presidential Authority', *Transition*, 30 June 1995, 35.
44 Author's interview, Almaty, Kazakstan, 10 June 1996.
45 Author's interviews in Almaty, Karaganda, Petropavlovsk, Zhambyl and Chimkent, summer 1996.
46 On the new Kazakstani constitution and the extension of President Nazarbaev's term in office.
47 Author's interview, Almaty, Kazakstan, 13 September 1996.
48 Author's interview, Petropavlovsk, Kazakstan, 6 August 1996.
49 Dave, 'A New Parliament', 35.
50 The Kazakstani capital was transferred from Almaty to Akmola (now renamed Astana) in June 1998.
51 Author's interview, Almaty, Kazakstan, 10 July 1996.
52 Assotsiatsya sotsiologov i politologov g.Almaty, *Social-Economic Situation of the Republic of Kazakstan during the Transitional Period* (English translation) (Almaty, 1996), 27–37.
53 Author's interviews, Kazakstan, summer 1996.
54 Author's interview, Petropavlovsk, Kazakstan, 5 August 1996.
55 Author's interviews, Kazakstan, summer 1996.
56 Assotsiatsya sotsiologov i politologov g.Almaty, *Public Opinion: Present Status, Problems, Outlooks* (Almaty, 1996), 35.
57 Author's interview, Almaty, Kazakstan, 10 September 1996.
58 Assotsiatsya sotsiologov i politologov g.Almaty, *Public Opinion*, 42.
59 Author's interviews, Almaty, Kazakstan, summer 1996.
60 American Legal Consortium, 'East Kazakstan's Green Party Rises from the Ashes of Chernobyl', *Central Asian NGO News*, June 1996, 1.
61 Informal conversations, Karaganda, Kazakstan, 24 June 1996.

Select bibliography

Akiner, S. *The Formation of Kazakh Identity: From Tribe to Nation-State* (London, 1995).

American Legal Consortium, 'East Kazakstan's Green Party Rises from the Ashes of Chernobyl', *Central Asian NGO News*, June 1996, 1.

Assotsiatsya sotsiologov i politologov g.Almaty, *Politicheskiye Partii i Dvizheniya (II)* (Almaty, 1996).

—— *Public Opinion: Present Status, Problems, Outlooks* (Almaty, 1996).

—— *Social-Economic Situation of the Republic of Kazakstan during the Transitional Period* (English translation) (Almaty, 1996).

Critchlow, J., 'Central Asia: How to Pick Up the Pieces?', in J. DeBardeleben and J. Hannigan (eds), *Environmental Security and Quality after Communism: Eastern Europe and the Soviet Successor States* (Oxford, 1995), 139–53.

Dave, B., 'A New Parliament Consolidates Its Presidential Authority', *Transition*, 30 June 1995, 33–7.

—— 'Kazakhstan Staggers under Its Nuclear Burden', *Transition*, 17 November 1997, 12–13, 72.

Davis, R. A. and Steiger, A. J., *Soviet Asia* (London, 1943).

Dawson, J. I., 'Anti-Nuclear Activism in the USSR and Its Successor States: A Surrogate for Nationalism?', *Environmental Politics*, 4 (1995), 441–66.

DeBardeleben, J., 'The New Politics in the USSR: The Case of the Environment', in M. Stewart (ed.), *The Soviet Environment: Problems, Policies and Politics* (Cambridge, 1992), 64–87.

Feshbach, M. and Friendly Jr, A., *Ecocide in the USSR: Health and Nature under Siege* (New York, 1992).

Glanyz, M. H., Rubinstein, A. Z. and Zonn, I., 'Tragedy in the Aral Sea Basin: Looking Back to Plan Ahead', in H. Malik (ed.), *Central Asia: Its Strategic Importance and Future Prospects* (New York, 1994), 160–94.

Goldman, M. I., 'Environmentalism and Nationalism: An Unlikely Twist in an Unlikely Direction', in Stewart (ed.), *The Soviet Environment: Problems, Policies and Politics* (Cambridge, 1992), 1–10.

Inhabitants of Egindybulak *Raion*, Karaganda *Oblast*, Kazak SSR, 'Letter to the Nevada-Semipalatinsk Movement', in H. B. Paksoy (ed.), *Central Asia Reader*, 177–83.

Kolarz, W., *Communism and Colonialism* (London, 1964).

McCauley, M., *Krushchev and the Development of Soviet Agriculture: The Virgin Land Programme, 1953–1964* (London, 1976).

Micklin, P. P., 'The Fate of "Sibiral": Soviet Water Politics in the Gorbachev Era', *Central Asian Survey*, 2 (1987), 67–88.

Micklin, P. P., 'Water Management in Soviet Central Asia: Problems and Prospects', in Stewart (ed.), *The Soviet Environment: Problems, Policies and Politics* (Cambridge, 1992), 88–114.

Olcott, M. B., *The Kazakhs* (Stanford, CA, 1987).

Smith, D. R., 'Kazakhstan', in P. R. Pryde (ed.), *Environmental Resources and Constraints in the Former Soviet Republics* (Oxford, 1995), 251–74.

United Nations Development Programme, *Aral Sea Basin NGO Directory: Environmental and Health Protection Non-governmental Organisations in Kazakhstan, Kyrgytstan, Tajikistan, Turkmenistan, Uzbekistan* (Tashkent, 1996).

Wolfson, Z., 'The Massive Degradation of the Ecosystems in the USSR', in Stewart (ed.), *The Soviet Environment: Problems, Policies and Politics* (Cambridge, 1992), 57–63.

Ziegler, C. E., 'Political Participation, Nationalism and Environmental Politics in the USSR', in Stewart (ed.), *The Soviet Environment: Problems, Policies and Politics* (Cambridge, 1992), 24–39.

9 Paths to ecofeminist activism

Life stories from the north-east of England

Niamh Moore

When we reach Karen's home it is easily identifiable, as her partner has described. Like Karen, the garden refuses to conform easily to the norms of middle-class conventions – unlike the gardens nearby with their neat herbaceous borders and carefully pruned front lawns. There's no gate, just a gap in the fence, almost invisible behind the full berry and currant bushes that border the garden. We push through into Karen's garden. It is a wild and unruly tangle, an eclectic mix of marjoram, azalea, lettuce, an overhanging wych elm tree, and more. I have come with my tape recorder to listen to Karen, again, and to two of her friends, Lesley and Jen. They are interested in ecofeminism and together they have recently formed a branch of the Women's Environmental Network (WEN) in the north-east of England and have been campaigning on various local issues. They are to tell me about their lives, their feminism, and their environmental activism, and how these are intertwined.

At a time when the women's movement is said to be in decline, and when activism is said to have abated from the heady levels of the 1970s, that increasing numbers of women are involved in ecofeminist activism should be the focus of some academic feminist interest. Feminism, however, has been reluctant to embrace ecofeminism, fearful of 'essentialism', of 'claims that women possess an essential nature – a biological connection or a spiritual affinity with nature that men do not'.[1] Such connections between women and nature have been used to relegate women to the home and child rearing, to domesticity, excluding women from the hard-earned gains of the public world of education and the workplace. Much recent writing from ecofeminist academics has concentrated on responding to this criticism.[2] My concern is the silence in this literature on the women activists who constitute this new movement, and the meanings that they make of ecofeminism – for activists not also involved in academia, essentialism is not a term that has much meaning.[3] My assumption has been that grounding debates about essentialism in the everyday lives of women, and the places where these lives are lived out, looking at how life experiences might lead to involvement in political activism, offers another path to understanding ecofeminism, potentially offering much more complex understandings of relationships between women and nature.[4]

Standing in Karen's garden even before I have switched on my tape recorder I am already confronted with questions about the relationships and borderlands between women, nature and culture. Gardening seems to have many meanings for Karen. It is self-expression, it is a document of a life story, it is ritual, it is political activism, it is evidence of the connection between the personal, political and spiritual in Karen's life. Later it transpires that a passion for gardening is a thread that connects the lives and politics of all three of these women. Their gardening is perhaps an attempt to make sense of this northern coastal landscape, with its decaying Victorian seaside grandeur, tall cliffs matched by tall chimneys of the chemical industry and its hinterland of farming and mining villages. Radically different landscapes co-exist side-by-side, dotted with Karen, Jen and Lesley's gardens, allotments and compost heaps.

Karen tells me that she planted the wych elm in her garden as a tiny twig when she moved here twenty years ago as a married woman with young children. She likes trees, they are important to her, and to our friendship – we had first met at a Peace Camp organized to protest the clear-cut logging of temperate rainforest in Clayoquot Sound on the west coast of Canada three years before. Since I last recorded some of her life story she has been thinking about trees from her childhood. She tells me about:

> the conker, the horse chestnut, which I planted in the garden, and it started to grow, I suppose I was about eight to ten, and over the winter I made a little, like, dunce's cap out of paper which I sewed up together to put round it to protect it from the cold, you know, and, then I fell out with my brother about something, and he threatened to go and break my tree down, you know, it obviously meant a lot to me, did this lovely little horse chestnut tree, so I was going to get him for this, you know, and my dad said oh, he won't do it, he won't do it, and he went out and he did do it, and I was very upset about it, and that was the last of my poor tree.

It is the first story that Karen chooses to tell me when I switch on my tape recorder. It is not where I wanted to start, but she has already mentioned the story earlier, and it seems important.

It is important, but it is later when I reflect on this that I begin to understand. Karen's brother's violence was not a random act. It was knowing and calculated. He knew the tree was important to Karen, more important than him perhaps? Karen's story reminds me of Carol Adams' contributions to ecofeminism. Her work looks at how violence against women and animals is intertwined, and provides a way for thinking through connections between the oppression of women and the destruction of the earth more generally. She has written about how in situations of domestic violence men often threaten, abuse and kill pets, both a violence in itself and a threat of further violence to women and children.[5]

Drawing on the work of Elizabeth Spelman, Carol Adams points out that 'somatophobia – hostility to the body – is symptomatic of sexism, racism, classism and speciesism, and demonstrates how hostility to despised and disenfranchised bodies, that is, those of animals, children, women, and nondominant men, becomes interwoven. To avoid somatophobia, feminist philosophy must take the connections between abuse of animals and abuse of women seriously'.[6] Further, she argues that environmental abuse is a form of somatophobia, 'that abuse of the earth is an expression of hatred of the earth's body'.[7] Such connections are difficult to make within a feminism that has been bent on severing any connections between women, animals and nature. A central strategy for many feminists has been the proclamation that any differences between women and men are the result of 'socialization', that gender differences are socially constructed and absolutely not biologically determined. This approach reproduces (sex/gender, female/male, nature/culture, body/mind, animal/human) dualisms by accepting the patriarchal devaluation of bodies and nature in the west, and neatly sidesteps the question of borderlands, between women's bodies (and men's), between animals' bodies and the earth's body, all regions where the influences of nature and culture ebb and flow. It's just not so easy to determine where, indeed if, nature ends and culture begins.

Carol Adams bases her work, in part, on her involvement with the refuge movement in the US, and the stories she has heard from battered women. These kinds of empirical connections between women and nature are not often represented in the literature on ecofeminism, but this is what interests me. I want to know how women experience connections between their own oppression and the domination of nature in their everyday lives, I want to know what life experiences bring women to ecofeminism, and what it means to them. It is through coming to their homes, to the places where they live, through the process of listening to, recording and ultimately writing about the lives of Karen, Jen and Lesley that I want to help to generate and create new texts of ecofeminism which will both add to and challenge the existing feminist and ecofeminist literatures.

Jen appears the most conventionally middle-class of these three women – there's no wild and unruly tangle before her house. She leads an apparently comfortable middle-class existence with her two children and her husband, but she tells a story where pain is only just beneath the surface. Rereading Betty Friedan's *The Feminine Mystique* thirty years or so after it was first published, Jen recognizes the relevance of Friedan's 'the problem that has no name' to her own life in the mid-nineties. Jen says 'to a certain extent I'm living that kind of life at the moment . . . and I've certainly got that ailment that has no name'. While she clearly states that 'becoming a mother was the thing which changed me most', her account of how motherhood led her to environmentalism and ecofeminism is not a straightforward account of maternal concern over the future of her children and grandchildren. Her first child nearly died, was ill for many years, and Jen experienced very serious post-natal depression. Once she had overcome the worst of her post-natal depression she found that she spent a lot of time talking to women who had similar experiences and through this she realized that 'the fact that I was totally isolated from a lot of women, that for a large part I was having this baby in a vacuum, all on my own, and that the pressures in that seemed very high, so I started thinking again about ideas of community and how, y'know, [we] really ought to be having children'.

The kinds of friends that Jen had before her experience of her post-natal depression were, as she described them, 'down-the-pub' friends – friends mainly from the Labour Party, friendships that were inadequate for this change in her life, friendships that could not change and accommodate her new experiences, as a parent at home, unable to go 'down-the-pub' so often, not least because of her depression. It is an experience that led her to rethink community, 'trying to think up a way . . . that would have been a rejection of "I'm a total failure as a mother, I've been depressed, I can't cope" . . . to do with a way of finding, of thinking, what kind of structures have prevented you from giving your kids a healthy childhood, what kinds of communities and societies would you be able to do that in'. The mining community that her grandparents have come from, that her mother no longer lives in, is where Jen often feels more at home. It is back to the mining villages of her grandparents that Jen goes with her first child, it is where she can begin to cope with her post-natal depression, where other people will help with the care of her child, where she feels less isolated. It is to her grandparents in this mining community that Jen went as a child, following her parents' divorce, when her claustrophobic relationship with her mother was overwhelming and her mother's nomadic lifestyle became too much. Her grandfather's phlegmatic personality provided the necessary calm from her mother's depression. As Jen says, 'sometimes I've tried to go back through my life and think about it and think, what were the essential things that was me, probably because sometimes I feel like a composite bit of all sorts, that I don't really have a history . . . what essential things did I really care about that I can relate to now?'

Having children has also reminded Jen of her own childhood, 'things became very important, which I can remember being important as a little girl, about y'know sort of, wild flowers and gardening and those things which get crowded out in your adult life, you know, through my children, I began to put back in centre stage'. For a while she contemplated moving back to this mining community, but now she describes the women's groups that she is involved in as her 'family of choice'. She has spent much of her life in the company of men, at Oxford, while she was in the Labour Party. Her experience of post-natal depression, however, has highlighted the necessity of female friendships. It was in the women's group that she met Karen and Lesley, and that they decided to set up a branch of the Women's Environmental Network, and this is where she currently focuses some of her energy.

For Karen, nature and animals were clearly part of her chosen community as a child. She tells how as a child, she felt that the cat was the only one who she could really talk to in her house. She tells me also about how horrified she was when she was told at school that animals did not go to heaven, and that she decided instantly that if animals didn't go to heaven then she didn't want to either. It seems significant though that after the story of her brother's destruction of the tree she had planted, there are no references to nature again until adulthood. Her interests shift from the town where she lived to the city of Liverpool, where her grandparents lived. From the age of 11 she is allowed to travel on her own to spend summer holidays with them. It is also at the age of 11 that she is sexually abused by her mother's employer, though she tells the two stories separately. I can't help but wonder if there is a connection, if part of the attraction of Liverpool then, is that the landscape of Liverpool's urban docklands where she wanders alone is safer for her than her own town. The abuse only stops at the age of 13 when she passes her scholarship and goes to grammar school. She says that she 'stopped it', that she became 'more interested in intellectual things', and became very studious. She says she doesn't think that she really suffered but that she 'didn't put any worth on her body', that she devalued herself. She decides that she wants to work in Liverpool, and that she wants to work with the poor in Liverpool. She goes to university although it is not expected of her. She becomes a social worker, she marries and moves to Sheffield, and eventually to the north-east; she is very involved in Labour politics, and becomes a local councillor. But everything changes when she starts a relationship with another woman. Shortly after this she goes to Greenham Common Peace Camp; she begins to identify as a lesbian and a feminist. She begins to make changes in her life. Her marriage ends. Her involvement in Labour politics tails off. She goes to Greenham Common Peace Camp every weekend for two years, and is part of a local Greenham support group. There are endless discussions about feminism and lesbianism. She says 'there has never really been anything to replace it'. Woven through Karen's account are stories about women in her family, her mother, her grandmother and her aunts, many of whom had strong friendships with other women. In trying to make sense of these changes in her life she looks back to these women's lives, and forward to her own daughter's, and to other young women, like myself.

Lesley's terraced house is cosy and welcoming. The air is aromatic. She has been burning basil in an oil burner to clear her head before I arrive. The kitchen where we sit and talk is full of interesting clutter, small bunches of herbs and flowers, wholemeal bread, a pot of homemade jam. It is a long way from the housing estate in Leeds where she grew up. When first talking to me she hardly mentions childhood and adolescence, she says little about her family, family friends or other generations of her family. When I ask her about this, it seems that as she sees little connection between her life then and now, she chooses to focus on the present, though eventually she tells me a little about her early life. Her father was an unskilled labourer, her mother had a series of part-time jobs as a seamstress. Her father was authoritarian. She says of her family 'we didn't do emotions'. Growing up she learnt that it was weak to display feelings, that she would be punished. As a child she felt she did not fit in, that she had been 'dropped in the wrong family', and she escaped as much as she could, hiking on the Yorkshire Moors and creating her own fantasy world, making up games and shows, and groups where she could be the leader. She created another world for herself. When she was older she escaped to university, but she feels it was an escape through her intellect only, not her whole person. In her late twenties, shortly after leaving university, she contracted tuberculosis. It was an experience that changed her life. She was very ill for a long time, and spent a year confined to bed. She became involved in alternative communities and, like Jen, tried to find ways of living differently. She moved to Hebden Bridge in Yorkshire and then to the north-east. She says that she often tells her life as if it began in 1970 after having TB, that she wiped out those parts of her history that seemed so incongruous to her. She says she has lived in a fantasy world for over forty years, and more recently she has been trying not to deny her earlier life, to accept those parts of her history. She sees her TB in part as a response to the way she was living, denying her body, her feelings, her passions. She says, 'I just want to be a part of anything that protects that physical thing from being completely trodden on, whether it's women, body, earth . . . The body is sacred, everything is sacred to me.' She feels that there was a lot of fragmentation in her life, and that her subsequent involvement in green politics and women's groups healed her some, and helped to bring bits of her together. She has been a Green Party councillor and has campaigned locally to improve people's lives. These politics enabled her to make connections, but to some extent those connections actually remained distant. She has spent a long time crusading and fighting for issues that are 'out there'. More recently she has felt the limits of this kind of political activity in 'freeing oneself' – this is not how she wants to live. She has also been involved with a village arts organization – it is an effective and fun way of presenting alternatives, and involving others in this process. She enjoys making costumes, singing and dancing. She says that in the past she used her creativity as a way of getting around issues, without having to confront them. This seems to be changing. She stills dances, and derives much pleasure from this. She has also recently completed a piece of visual artwork, entitled 'Chrysalis – she came out of the attic and danced and danced'. The title is so evocative for me, bringing together 'the mad woman in the attic' and Emma Goldman's 'If I can't dance, then I don't want to be part of your revolution.' The one thing that remains constant in her life is her work as a gardener. She counts the old gardeners as her friends; they give her plants, and are into the 'natural recycling that people have done for years'.

Karen, Jen and Lesley rewrite their lives from their own understandings of their lives rather than what others tell them about their lives. Their stories challenge the marginalization of their experiences as women, and the naming and silencing of these experiences by others. Their stories make women central in their lives, shift interpretation to what they have learnt from other women rather than from the dominant cultural messages, as Jen puts it, 'learning from other women that you are not the only one'. It is not enough to start this project from this generation. In their telling of their life stories, Karen and Jen, and to a lesser extent Lesley, rewrite their genealogies,[8] albeit in ways that vary for each woman. These genealogies are analogous to the attempts of ecofeminists to reclaim a matrifocal or matrilineal era in history for women. Linda Vance calls these 'conjectural histories', which she sees as having an entirely different aim from conventional history which seeks to represent some objective account of the past. Vance sees that 'many ecofeminists . . . find greater satisfaction in stories of the past that acknowledge the intangible, the magical. And many of us want more from history than a cautionary reminder of what we already know. We want inspiration and alternative ways of knowing.'[9]

Rewriting genealogy, for Karen and Jen in particular, provides a way of finding continuity in life, a way of explaining the many different interests and directions that their lives have taken. It is a genealogy that is based on the creation of 'communities of choice' rather than the necessities of 'families of origin'. In particular rewriting genealogy seems a way of including nature in their 'chosen community' and in this process transforming their own lives and the places where their lives are lived out.[10] Karen, Jen and Lesley contest dominant (Christian) stories that have come from other human bodies, stories which vary from many other creation myths throughout the world which tell creation stories of connection with, rather than separation from, the earth. This retelling of where they come from corresponds with part of what Linda Vance sees as the 'project of ecofeminism', which is:

> understanding, interpreting, describing and envisaging a past, a present, and a future, all with an intentional consciousness of the ways in which the oppression of women and the exploitation of nature are intertwined. Without an appreciation of the past, we don't know where we've come from, without knowledge of the present, we can't know who we are. And, most critically, without a vision of the future, we can't move forward.[11]

This retelling of these stories is an important component of their politics. These stories are about different ways of knowing, ways that aren't always legitimated by the rest of the world. While Karen, Jen and Lesley have all been to university, and all have degrees, they are all critical of what they have learnt there, what counted as knowledge, and what didn't. Lesley clearly sees that there are other ways of knowing:

> you've got it in you to become every bit as individual as a plant that grows that's an individual and has its own smell and colour and every-

> thing . . . and you've got it all in you before you go to any school . . . I mean it's nice to be able to read and write, but, I mean, it's all there, and I just wonder how many people ever flower you know really, are allowed to really flower, and then if everybody did, there wouldn't be this confrontational thing, you'd all be so interested in being yourself, there's somebody else that's not like me, I must get to know them . . . you just wonder how many things are stamped on, like how many natural things, never get going.

Both Karen and Jen have returned as mature students to university to take courses in Women's Studies. Ten years ago Karen took an MA and Jen is currently completing an Open University degree which involves some Women's Studies courses. Jen realized that in the courses she took during her first degree:

> I always did a certain amount of, if you like, translating, working on these courses and they sort of referred to me at a distance, I had to actually place myself in relation to them, I wasn't actually there, and when I started reading things that were by and about women, I was startled at, by how direct that was, by how directly that affected me, that there were these things that I didn't actually have to translate, and that was exciting.

While it is on her course that Jen hears of ecofeminism, it's a passing reference, there's one short chapter in one of the textbooks, that's all. Ecofeminism is notable for its absence in British feminism. It is when reading Karen's dissertation for her MA in Women's Studies that it all becomes clearer.[12] I have in my hand the essay that I have always wanted to read – a genealogy in its own way of feminism in the eighties in Britain. It is an argument about the importance of Greenham Common for feminist politics. It is a response to feminist critics of Greenham, based on representations of Greenham Common women in the newsletters from the camp and on the practice and philosophy of the camp, which she identifies as ecofeminist. Karen suggests that 'the Greenham philosophy has been misunderstood by the radical and socialist strands of feminism. Maybe this is partly because ecofeminism has developed through practice rather than by relying on literary exegesis, but it is also because other feminists have not studied the history behind the philosophy.' She identifies this philosophy as based on non violence, consciousness-raising and spirituality. I have always wondered that more hasn't been made of Greenham in academic women's studies circles in Britain, but I begin to understand, as Karen must too, given that her thesis was failed by her examining board. I wonder why and how such a significant part of the consciousness-raising of many women's lives can be absent from the history of the women's movement in Britain.[13] It seems no coincidence that Karen,

Jen and Lesley have all been involved in various ways visiting and supporting the women at Greenham Common. What is clear from their stories is the profound need for more stories of the emergence of feminism in Britain.

While Karen, Jen and Lesley all have a considerable history of involvement in traditional party politics, significantly none of the women is any longer involved in this kind of politics in the same way, although their concern about political issues has far from abated. Karen and Lesley were councillors, Karen for the Labour Party and Lesley for the Green Party, and Jen, while never an elected councillor, was also heavily involved in local Labour politics. While the Green Party would undoubtedly see itself as an entirely different kind of political party from Labour, the party that is 'neither left nor right but ahead', what Karen, Jen and Lesley have to say about their disaffection with these parties is quite similar. Ultimately none of these parties could adequately encompass all of the issues that they wanted to include. As Jen says now of what she is looking for, 'coming out of conventional politics, I'm looking for a politics that addresses a lot more of what I am and what's important in my life than having a political category into which you fit'. This is particularly important for Lesley who is becoming involved in politics again after a six year break while she was involved in an abusive relationship. Joining the women's group and WEN have helped to give her the strength to leave her relationship. She sees parallels between her experiences, saying:

> and that's a bit like how I felt drained in those sort of politics, is a bit like I feel in the wrong relationship. I can see a sort of parallel there as well, you stay in something, out of a sort of sense of duty or something, and you feel you've got to do your best, do it conscientiously, but it's actually doing you no good at all, so I do think that coming out of that and coming into what I call my more positive groups is also a part of establishing who I really am and what I really want and who I would draw to me, that's all the same stuff as well.

She sees that 'a lot of conventional politics is about having some sort of power, over the people, to make decisions for them', and she is sceptical of what this kind of politics can achieve. She tells me about a road that has been built through the village. The previous road used to curve around the shops, and meant that traffic had to drive slowly through the village. The new road is very fast, it has effectively cut the village in two, and changed the pace of the community. It was a decision made by the county council with little consultation with local people. Two people have since been killed crossing the road. 'Real power', she says, 'is about trying to understand yourself . . . and then it being a bit easier to understand other people and work with other people . . . and it slows everything down, that sort of process, so you can't just go out and build instant motorways because you are too busy thinking, "would that hurt anyone, what would they think?" '

One of the attractions of ecofeminism for Karen, Jen and Lesley is precisely that it connects everything up, allows for the inclusion of every aspect of their lives:

> Greenham brought so many things together for me, it did bring socialism and working on behalf of the oppressed people and being on the side of the oppressed together and lesbianism, and women's interests, and ecofeminism, though I wouldn't have called it that . . . brought it all together – and that was one of the really important things for me. (Karen)

> I think that was the thing about working in these women's groups, it's such a whole thing, it includes everything . . . every aspect of your life. (Lesley)

> . . . it's no good going to church on a Sunday and then going home and putting weedkiller on the garden, it, it just doesn't fit for me you know. (Lesley)

> . . . and ecofeminism, when I came across the idea of ecofeminism, it was surprising to me because it seems to bring together, sort of lots of separate things, I felt that there were lots of . . . bizarrely connected between me making compost in the garden and my politics and my spiritual interests, but I never actually saw anything written down on paper. (Jen)

In trying to make meaning of their lives and politics, Karen, Jen and Lesley challenge prevailing discourses within feminism of multiple and fragmented identities. Susan Griffin suggests:

> So, a problem raised by poststructuralist language philosophy might find the beginnings of an answer in ecofeminism . . . But where poststructuralist feminism chooses language as both its site and its focus, ecofeminism chooses ecological processes which make up the earth and the atmosphere as a source of knowledge.[14]

Karen, Jen and Lesley live in the north-east of England, the place that connects their politics and friendship. Now the three of them choose to do politics in the context of the Women's Environmental Network Group (WEN) which they have set up. This group has emerged in part out of another women's group that all three were involved in. The WEN Group has been involved in a number of issues, a local bypass being built, a campaign to open a local disused railway line. Significantly though, the most popular issue has been the recording of local sacred sites, which was part of a national WEN survey. For Lesley one of the important aspects of the sacred sites campaign was the pleasure of doing something that she defined as political, but which didn't have to involve traditional methods of doing politics which she sees as mainly confrontational. She juxtaposed this with the railways campaign, which involved many meetings with other local interest groups, where almost all the other participants were men, and levels of confrontation were high. One important aspect of the sacred sites campaign was the way in which it provided a way of 're-visioning a known landscape'.[15] Jen says:

> I've always had a very negative environmental image of the area . . . and I still always worry about living here in R—, I don't like the pollution, worry about the children, and I constantly want to move out and things, and I thought it was really nice to have something celebratory about the environment locally and we try to tie that to women as much as possible, and we certainly, well I did definitely feel differently about the whole place, after we started, we definitely had a different attitude . . . maybe we had a longer political perspective on this piece of land . . . and the chemistry industry is very recent. It had never occurred to me that my grandparents . . . used to complain about me being disillusioned about living in R— and wanting to move, and disliking it intensely, that the chemical industry wasn't there, that the R— they knew was totally different.

Through the sacred sites campaign Jen gains a new relationship with the surrounding area. Linda Vance has pointed out how women lack the same history of relationships with land which men can draw upon. She sees this history as necessary to give depth and meaning to women's lives, and thus highlights an important site for future ecofeminist research. Further she suggests that, 'Important for women is our need to know our own experience with nature more fully. Many of us are separated from everything that is natural – including our own bodies – by centuries of patriarchal domination.'[16] There has recently been much feminist writing on 'the body', though the overwhelming emphasis is on how 'the body' has been culturally constructed. Carol Bigwood critiques this work, believing that 'if we reduce the body to a purely cultural phenomenon, then we perpetuate the deep modern alienation of our human body from nature . . . The way to deconstruct the nature/culture dichotomy is not to make everything cultural.'[17] This 'body' of much feminist theorizing is, paradoxically, often disembodied and dislocated, suspended in a very postmodern space. As Susan Bordo says, 'the postmodern body is no body at all'.[18] While ecofeminism (and feminism) parallel postmodernism in their calls for attention to the particular rather than the universal, to 'the local' as the site of politics, as Elspeth Probyn recognizes it is 'an unspecified local' that has become 'the site for an unnamed politics'. Probyn suggests that 'a feminist reworking of these metaphors may bring them down to earth', though there is little to suggest that feminism has managed this.[19] Ecofeminism, such as that of Karen, Jen and Lesley, calls for a 'politics of location', but one which recognizes that we are situated not only by race, gender, ability, sexuality – but also by our relationship with nature and with the physical landscape. Such a politics can avoid both relativism and essentialism, if location is not just understood as a metaphor that is inscribed on the body.

I suggest that an ecofeminist reworking of the metaphor would acknowledge the role of 'real' places in our lives, that ecofeminism would propose an 'understanding of self based on a conversation with the land'.[20] Karen, Jen and Lesley seem to suggest that ecofeminism offers a politics which is deeply committed to activism, but also a politics that offers strength through a spirituality that is a source of empowerment. Karen plants another tree, finally:

> There's a big wych elm out the front, which is as high as the house now, and when I planted it, I suppose it's twenty years ago, it was just a little twig, you know a little divided twig, like that, and it's just amazing how big and strong it's grown, it really, you know, there really is something special about them.

For Karen, Jen and Lesley their changed relationship with the land around where they live, with the villages that are part of the landscape, and the people who live there, is the basis for their politics.

When I ask Lesley at the end if there is anything else she wants to say, she tells me of the importance of the area where she lives, 'just one thing that occurred to me, getting involved in different areas, like this particular area has a lot of work, there's a lot of work to be done here, I don't know if I lived in London or Leeds whether I would be so involved, this way it feels, kind of, I could do something'. Karen, Jen and Lesley draw on their own life stories and the places where these lives have been lived out to rewrite their genealogies, creating communities of choice that include nature. For Karen, Jen and Lesley, ecofeminist politics is about finding the place, both physical and spiritual, to do politics, and to do politics differently.

Notes

I would like to acknowledge the following: Karen, Jen and Lesley (these are pseudonyms), for sharing their experiences of ecofeminism and their stories of their lives with me; Sue Wright for sharing her knowledge feministly, for walks and talks on the Sussex Downs, and for wonderfully detailed comments and suggestions on an earlier draft; Al Thomson and Dorothy Sheridan for the pleasure of taking part in the Life History Certificate at the University of Sussex, and to all the students, especially Lorraine Sitzia and Vijay Reddy. Thanks also to Peter Dickens, all at the feminist theory reading group, Chris Abuk, Margaretta Jolly, Lucy Ford and Jacinta French, who read and commented on earlier drafts; and to Sharon Krummel.

Interview Dates: Karen: 8 February 1996 and 19 June 1996; Jen: 20 June 1996; Lesley: 20 June 1996.

1 J. Birkeland, 'Ecofeminism: Linking Theory and Practice', in G. Gaard (ed.), *Ecofeminism: Women, Animals, Nature* (Philadelphia, PA, 1993), 22.

2 Examples I have in mind include V. Plumwood, *Feminism and the Mastery of Nature* (London, 1993); K. J. Warren with B. Wells-Howe (eds), *Ecological Feminism* (London, 1994); C. J. Cuomo, *Feminism and Ecological Communities: An Ethic of Flourishing* (London, 1997); N. Sturgeon, *Ecofeminist Natures: Race, Gender, Feminist Theory and Political Action* (London, 1997).

3 Noël Sturgeon's work is a rare instance of a recent book on ecofeminism that does discuss activism, and is to be commended for this. There has been some discussion of the lives of 'well-known' ecofeminists – see for example, on Rachel Carson, P. Hynes, *The Recurring Silent Spring* (Oxford, 1989) and L. Lear, *Rachel Carson: Witness for Nature* (London, 1998); on Petra Kelly, S. Parkin, *The Life and Death of Petra Kelly* (London, 1994). A. W. Garland's *Women Activists: Challenging the Abuse of Power* (New York, 1988) is one of the few books to include life stories of some less well-known women environmental activists. What little there is written more generally about why women become involved in environmental activism rarely gets beyond very simplistic accounts of maternalism, nimby-ism, or in the case of discussions of the Chipko movement, immediate subsistence needs, as explanations for women's activism.

4 Here I draw on the work of Jim Cheney who begins the process of developing an 'ethic of bioregional narrative' which he sees as a way of 'grounding narrative without essentializing the self'. J. Cheney, 'Nature/Theory/Difference: Ecofeminism and the Reconstruction of Environmental Ethics', in K. J. Warren with B. Wells-Howe (eds), *Ecological Feminism* (London, 1994), 158–78. I also draw on Carol Bigwood's work on 'an environmental ethic of place' in C. Bigwood, *Earthmuse: Feminism, Nature, and Art* (Philadelphia, PA, 1993). Other ecofeminist writers who have influenced me here are S. Lahar, 'Roots: Rejoining Natural and Social History', and L. Vance, 'Ecofeminism and the Politics of Reality', both in Gaard (ed.), *Ecofeminism*, 91–117, 118–45; and T. T. Williams, *Refuge: An Unnatural History of Family and Place* (New York, 1991).

5 C. J. Adams, 'Bringing Peace Home: A Feminist Philosophical Perspective on the Abuse of Women, Children and Pet Animals', in C. J. Adams, *Neither Man nor Beast: Feminism and the Defense of Animals* (New York, 1994), 144–61, and 'Woman-Battering and Harm to Animals', in C. J. Adams and J. Donovan (eds), *Animals and Women: Feminist Theoretical Explorations* (Durham, NC, and London, 1995), 55–84.

6 Adams, 'Bringing Peace Home', 145.

7 Ibid. 157.

8 I use genealogy here in the Foucauldian sense. I also draw on Terry Tempest Williams' *Refuge*, and her exploration of genealogy.

9 Vance, 'Ecofeminism', 129–30.

10 See L. Gruen, 'Towards an Ecofeminist Moral Epistemology', in K. J. Warren with B. Wells-Howe (eds), *Ecological Feminism*, 120–38, on what community might mean for ecofeminism.

11 Vance, 'Ecofeminism', 126.

12 Manuscript in possession of the author.

13 S. Roseneil's *Disarming Patriarchy: Feminism and Political Action at Greenham* (Buckingham, 1995) is the first full-length sociological analysis of the Greenham Camp.

14 S. Griffin, 'Ecofeminism and Meaning', in K. J. Warren with N. Erkal (eds), *Ecofeminism: Women, Culture, Nature* (Bloomington, IN, 1997), 216–17.

15 With thanks to Sue Wright for this expression.

16 Vance, 'Ecofeminism', 140.

17 Bigwood, *Earthmuse*, 10.

18 S. Bordo, 'Feminism, Postmodernism and Gender-Scepticism', in L. Nicholson (ed.), *Feminism/Postmodernism* (London, 1990), 145.

19 E. Probyn, 'Travels in the Postmodern: Making Sense of the Local', in Nicholson (ed.), *Feminism/ Postmodernism*, 177.
20 R. Raglon, 'Literature and Environmental Thought: Re-establishing Connections with the World', *Alternatives*, 17(4) (1995), 30–1.

Select bibliography

Adams, C. J., *Neither Man nor Beast: Feminism and the Defense of Animals* (New York, 1994).
Adams, C. J. and Donovan, J. (eds), *Animals and Women: Feminist Theoretical Explorations* (Durham, NC, and London, 1995).
Bigwood, C., *Earthmuse: Feminism, Nature, and Art* (Philadelphia, PA, 1993).
Birkeland, J., 'Ecofeminism: Linking Theory and Practice', in G. Gaard (ed.), *Ecofeminism: Women, Animals, Nature* (Philadelphia, PA, 1993), 13–59.
Bordo, S., 'Feminism, Postmodernism and Gender-Scepticism', in L. Nicholson (ed.), *Feminism/Postmodernism* (London, 1990), 133–56.
Cheney, J., 'Nature/Theory/Difference: Ecofeminism and the Reconstruction of Environmental Ethics', in K. Warren with B. Wells-Howe (eds), *Ecological Feminism* (London, 1994), 158–78.
Cuomo, C. J., *Feminism and Ecological Communities: An Ethic of Flourishing* (London, 1997).
Gaard, G. (ed.), *Ecofeminism: Women, Animals, Nature* (Philadelphia, PA, 1993).
Garland, A. W., *Women Activists: Challenging the Abuse of Power* (New York, 1988).
Griffin, S., 'Ecofeminism and Meaning', in K. J. Warren with N. Erkal (eds), *Ecofeminism: Women, Culture, Nature* (Bloomington, IN, 1997), 213–26.
Gruen, L., 'Towards an Ecofeminist Moral Epistemology', in K. J. Warren with B. Wells-Howe (eds), *Ecological Feminism* (London, 1994), 120–38.
Hynes, P., *The Recurring Silent Spring* (Oxford, 1989).
Lahar, S., 'Roots: Rejoining Natural and Social History', in Gaard (ed.), *Ecofeminism: Women, Animals, Nature* (Philadelphia, PA, 1993), 91–117.
Lear, L., *Rachel Carson: Witness for Nature* (London, 1998).
Parkin, S., *The Life and Death of Petra Kelly* (London, 1994).
Plumwood, V., *Feminism and the Mastery of Nature* (London, 1993).
Probyn, E., 'Travels in the Postmodern: Making Sense of the Local', in Nicholson (ed.), *Feminism/ Postmodernism*, 176–89.
Raglon, R., 'Literature and Environmental Thought: Re-establishing Connections with the World', *Alternatives*, 17(4) (1995), 28–35.
Roseneil, S., *Disarming Patriarchy: Feminism and Political Action at Greenham* (Buckingham, 1995).
Sturgeon, N., *Ecofeminist Natures: Race, Gender, Feminist Theory and Political Action* (London, 1997).
Vance, L., 'Ecofeminism and the Politics of Reality', in Gaard (ed.), *Ecofeminism: Women, Animals, Nature* (Philadelphia, PA, 1993), 118–45.
Warren, K. J. with Wells-Howe, B. (eds), *Ecological Feminism* (London, 1994).
Warren, K. J. with Erkal, N. (eds), *Ecofeminism: Women, Culture, Nature* (Bloomington, IN, 1997).
Williams, T. T., *Refuge: An Unnatural History of Family and Place* (New York, 1991).

10 Pathways to the Amazon

British campaigners in the Brazilian rainforest

Andréa Zhouri

> The privilege of standing above cultural particularism, of aspiring to the universalist power that speaks for humanity, for the universal experiences of love, work, death, and so on, is a privilege invented by a totalizing western liberalism.[1]

Introduction

Historically, western anthropologists have travelled to places like Africa and the Amazon to study 'native' peoples – the 'primitive'. My own route to the Amazon has taken me in the opposite direction to that of the classic anthropological tradition: as a *western* Brazilian, I have found my 'natives' in Western Europe.[2]

During the last three decades environmental movements have become one of the most significant political forces in the west. Many critical studies have analysed and classified environmental movements and organizations in the realm of the so-called new social movements.[3] They are discussed in terms of their ideology, organizational structure, political style, accountability, performance, base of support, and even the motivations of their supporters. However, although recognizing the importance of how such movements brought new perspectives on the qualitative aspects of life, and perhaps echoing their increased professionalization in recent years, there has been little account of the individual experiences of activists inside environmental organizations. Who are the campaigners of the NGOs (Non-Governmental Organizations)?[4] How do they relate their own development to the issues they take up and the organizations they work for? And what are the implications of their attitudes for the practices of NGOs? These are some of the general background questions underlying the present discussion which focuses in particular on NGO campaigning for the Amazon rainforest.

The Amazon is one of the most powerful symbols of global environmentalism. It is constituted by social, political and historical places and spaces whose images are highly contested by different groups not only at local and national levels, but also in the global realm. Thus, for a local rubber tapper, a riverine *caboclo* or an Indian, the Amazon might represent a living resource for everyday life. For a Brazilian general it might be a frontier to defend. A multinational company might see timber to be exported, while scientists see the biggest biodiversity ecosystem to be investigated, and anthropologists see cultural diversity. Travellers seek adventure and 'joy in nature', whereas environmentalists value the forest as a vital natural resource and a home to people whose way of life is threatened by transnational capital.

Consequently, by cutting across nation state and cultural boundaries, the Amazon appears to be a transnational political space around which issues and dilemmas constitute themselves. At the same time as it exposes conflicts between different groups in local, national and global dimensions, it is also a territory in which a transnational 'imagined community'[5] is constructed and contested, through the forming of alliances: for example, between a variety of transnational NGOs, as well as between NGOs and forest dwellers. That is why it is such a powerful symbol of contemporary global environmentalism; and why I have researched the engagement of British-based NGOs with the Brazilian Amazon, its people and their counterparts in Brazil.

Commentators from a wide range of intellectual perspectives have discussed the processes of globalization not only in economic terms, but also in its cultural and social aspects.[6] Environmental crises, and the awareness and responses these have generated in the last twenty-five years, have attracted similar analysis.[7] Nevertheless, in contrast to the research on which this chapter is based, their approaches have not substantively and comprehensively considered the ways in which NGO campaigners, particularly environmentalists, both express and experience these 'global' processes.

In a broad sense, Giddens refers to globalization as 'action at a distance'.[8] Its intensification over recent years is related to the emergence of means of instantaneous global communication and mass transportation. Thus, within the new scope of time and space, novel forms of agency and political practices emerge, cutting across nation state boundaries and calling the latter into question. Time and space compression alongside the aspect of 'disaffection with orthodox political mechanisms' give rise to transnational agents in environmental, feminist, ethnic and peace movements/organizations.[9] However, global in their scope, these organizations and movements still express particular manifestations circumscribed by, among other factors, historical and cultural national legacies.

The Brazilian Amazon rainforest has been at the core of the actions of many groups world-wide. British-based groups have been among the most active in that region. Many important issues can be raised regarding 'global' environmental groups' actions in the Amazon. For instance, it is important to consider the ways in which global, national and local actors relate to each other and the power relations and conflicts resulting from their joint actions. Although I will not discuss all the aspects involved in this issue within the limits of this chapter, it is important to notice that from a nationalist Brazilian perspective, transnational NGOs are viewed with suspicion by many groups, especially local economic and political elites, and the military. They offer accusations of eco-imperialism and romanticism, and are suspicious that these groups act in the name of powerful economic interests from developed countries. NGO campaigners are aware of these perceptions and thus the fact that interviews were carried out by a Brazilian researcher certainly affected many of the attitudes, questions and answers to the issues raised. The Brazilian allegations of eco-imperialism and romanticism appear to be the main reference against which the discourses of the British campaigners are produced in the course of my interviews.

Over three years I carried out fieldwork among different sectors of British society, each with some involvement with the Brazilian Amazon. Sixty-seven interviews were conducted with campaigners, journalists, film-makers, researchers, members of the Timber Trade Federation and the cosmetic company The Body Shop. In recording the testimonies of campaigners I attempted to follow a life story form, but in doing so encountered difficulties.[10] These stemmed from a combination of: the images the campaigners think that Brazilians have of them; the allegations of emotionality against environmentalists always levelled by the private sector and governments whom they oppose; and the fact that as political actors and intellectuals themselves they do not regard their subjective or personal feelings and experiences as an important perspective on the type of work they do. Nevertheless, with further questioning, important clues did emerge and it is possible to trace some aspects of their personal trajectories in relation to: the issues they take up and the organizations they work for; how they built up their engagement with Amazonian issues; how they relate themselves as British political actors to the local context and people; and how they justify their actions.

In the following pages, I will identify and analyse the main tendencies in the 'British-Amazonian' NGO scene, based on the testimonies of campaigners. I shall specify the discursive elements that define the different perspectives; that is, the regularities composing each discursive tendency, with focus on the personal and social trajectories that lead the campaigners to engage with them, therefore producing a particular form of discourse and relationship to the Amazon.[11]

The tension constitutive of the Amazon as a field of communication and struggle between different positions within Brazil, Britain and the transnational realm can also be observed among campaigners. Following the Bruntdland report and its concept of 'sustainable development' in the late 1980s, a stance reinforced by the Earth Summit in 1992, a policy of incorporating social and economic dimensions into the 'environmental arena', as well as an environmental consideration into the economic agenda, took place through several governmental and non-governmental initiatives. However, this articulation remains an unsettled and ongoing dispute involving different social and political positions, in as much as the concept of 'sustainable development' itself corresponds to vague, shifting and contentious meanings.[12] A case in point is the fact that, in terms of the NGO transnational co-operation, issues of international inequality, trade and debt summoned the attention of 'Northern' environmental groups, and has also underlined their forest campaigns and their relationship to 'Southern' NGOs.[13] Therefore, the voicing of, and tension between, concerns with issues of 'biodiversity' and 'social justice' are constitutive of the campaigners' discourses about the Amazon. Nevertheless, such configuration assumes different forms, and stresses different meanings, according to the campaigners' personal and social trajectories.

By way of a graphic visualization, I want to delineate three tendencies among British campaigners for the Amazon: *trees*, *people*, and *trees and people*. 'Trees' signifies those campaigners that lay stress on environment/biodiversity concerns, 'people' stands for those that emphasize development/social justice issues, while 'trees and people' signifies the synthesis of the two former tendencies. The words 'trees' and 'people' are used in a metaphorical sense, as though, on the one hand, encapsulating concerns with the conservation, preservation, protection and sustainable uses of 'the environment' and, on the other hand, encapsulating issues of social justice, development and human rights. Furthermore, I speak of predominant *tendencies* among campaigners to highlight the heuristic and flexible nature of such categories and modes of classification, beyond the evident differences between NGOs and the actual and often irreducible complexities of the issues at stake. Therefore, the idea of a tendency among campaigners suggests that, in actual fact, there is a great deal of interplay, communication and tension between them, since, broadly speaking, most campaigners have acknowledged the entanglement of environmental and social justice issues in their campaigning activities.

Different routes/roots, different views/outcomes

The tendencies towards 'trees', 'trees and people' or 'people' will be here analysed *vis-à-vis* the interviewees' experiences of displacement, the growth of their political awareness, their involvement with the Amazon and their justifications for political campaigning.

A common element among those interviewed was that they all experienced travelling and displacement as part of their political engagement. For many, this 'displacement' started very early as part of their family experience and upbringing: in some cases a parent was a diplomat or military personnel, a company employee in Africa and Latin America, and there were those who moved around England or into England from Scotland, therefore perceiving themselves as migrants or even 'internationalists'. Nevertheless, they differ from British travellers of past centuries – colonial bureaucrats, explorers, geographers, scientists and anthropologists – who helped to install in the British imagination a fascination with 'exotic' nature and cultures while contributing to the expansion of the British empire and capitalism world-wide.[14] Unlike these previous representatives of the interests of national entities, NGO campaigners – as transnational political actors – represent a new relationship between citizenship and nation states, putting the latter into question by acting and intervening in ways that cut across their boundaries.[15]

Instantaneous global communication and mass transportation have made distances 'shorter', time and space have become compressed, and contact with different cultures now shapes personal experience of the world in a global way. Of course, such 'global' experiences require some preconditions in the form of financial means, access to new technology and linguistic skills. Certainly environmental and human rights agents share this 'global' experience. Thus remote areas have become closer and interlinked just as 'the exotic' has become familiar. However, this is not to say that environmentalists and human rights advocates all hold the same homogeneous image or understanding of the world. Neither is it to say that the intensification of contact implies a better understanding of and communication with 'the other'. Although sharing basic principles, there are different motivations for travel, different routes, different interactions with otherness and different forms of political engagement. It follows, as a consequence, that particular actors have different experiences and images, for instance the Amazon rainforest, with related meanings they create and reproduce within the universe of British-based NGOs.

'Being there', having practical or direct experience, is one important component of all testimonies.[16] Even though 'being there' lends authority to the political actor embarking on a globalized campaign, direct contact with the Amazon can lead to creation of differential discursive practices and varying political experiences.

Trees

For the group of interviewees with a strictly environmental-biodiversity interest, 'being there' assumes a scientific-political meaning. It is the research practice that supports their views and actions on a professional level. The campaigners who exemplify this tendency are found in the major environmental organizations, namely World Wide Fund for Nature (WWF), Friends of the Earth (FOE) and Greenpeace, regardless of the specific political and structural orientations of each group.

With the increasing professionalization of NGO work since the early 1980s, there has been an inclination towards the recruitment of forest campaigners with expertise in forestry, ecology, biology, geography, botany and related areas. A brief profile of campaigners reveals that the majority of them are young, generally in their late twenties and thirties, and have had academic training in the above mentioned areas of expertise. Although they have travelled to forested areas in Africa or Latin America as part of their fieldwork activities – 'field trips' to tropical forests – they generally have no personal experience of living in 'Third-world' countries or in tropical forest regions.

Significantly, their professional qualifications were the first point they made when I asked them: 'tell me about yourself'. This is what I observed in the testimonies of campaigners who were working for Greenpeace, FOE and also WWF at the time of the interviews. For example:

> I started doing a degree in Ecology, at Leeds University . . . and I have always been involved in NGO work . . . there is an organization called British Trust for Conservation Volunteers where you spend weekends or a week in the country helping out – *working on conservation issues*, whether it is building a stone wall or working in the forest or – *fieldwork really*. Yeah – since I was at school, I've been involved in that type of project . . . I was also interested in biology and geography. But *I wasn't interested in the pure science, really. I was interested on the link between that and what was happening today, what was happening in the environment* . . . [to] learn something that was relevant to the environmental problems you could see around. *So, eventually I wanted to work in a job or position where I found my decisions . . . would have a direct effect on stopping those environmental abuses.* (Greenpeace campaigner, female in her thirties)[17]

Similar impetus can be seen in a FOE campaigner:

> I was brought up all over the world because my father always moved around. I did a geography degree, and I did green politics and environmental issues. And I decided when I was a student that I wanted to work at Friends of the Earth. So, I graduated in 1990 and came straight to Friends of the Earth and I worked in the Rainforest team as a volunteer . . . Geography has always interested me: different parts of the world. So, I was always very aware of having a job which didn't just done [*sic*] you money and you went home at the end of the day and forgot about it. I always wanted to do something that I felt I was very very into personally and not just professionally. *My main influence was my tutor at the university.* I hated him [laughing]. It is very odd that he was the one who influenced me to go this way, but he did. He came up with this Marxist thing, he was very left wing and used to put his own politics to us. Which we didn't like. *We were very sort of apolitical, if you like. I actually found the geographical side of it: weather systems, and how pollution affects weather systems and climate . . . that really interested me. We have such a strong influence on how the world, the natural world is evolving and it's damaging* [*sic*] . . . *I just sort of realized that was what I wanted to do.* (Friends of the Earth campaigner, female in her twenties)

And another, from WWF:

> I've been at WWF for 8 years. I came to WWF from university. I did my MSc in Oxford in Forestry and Land Use. And then, I went to Forestry Conservation and initially I was working on field projects, particularly in West Africa, in Nigeria and Cameroon. And increasingly I've been working on policy, forest policy related to government and industry. (WWF campaigner, male in his thirties)

In these cases, it is the possibility of practising an 'applied science' and influencing the policy making related to forest issues that constitutes the basis or the motivating force behind their engagements. They initiated their professional careers at a time when campaigning was already run on a professional basis by NGOs. Although campaigning about the Amazon, the importance these campaigners attribute to 'being there' has not necessarily led them to the Amazon. For this specific group, Africa is the place mentioned as their first personal contact with a forest. It is referred to in the previous quotation from the WWF campaigner, and also in the following cases:

> I have been to Africa, but not to any rainforests. But obviously I would like to go. It would be interesting. (FOE campaigner)

> I've been enjoying travelling on my own and with friends, just because *I am very interested about what is happening in life around the world.* You know, I am curious and – in particular with Africa, I was very interested. *I have got this fascination with Africa. Reaching various places and wildlife, nature in Africa. And I felt it was very important to be . . . to see these places, to visit and have knowledge and experience behind me. I've travelled a lot with Greenpeace as well. But it is nice to travel and to be involved in something on the ground.* (Greenpeace campaigner)

In terms of wildlife, and also as a personal interest, it seems that Africa (and Asia, but less so Latin America) plays an important role as a topos in the British cultural imagination, as well as being a popular trope in western literature and colonial texts, particularly the travel genre of which the classic example is Joseph Conrad's *Heart of Darkness.*[18] An interest in exotic cultures and wildlife – two concepts which are always interlinked – underlies both imperialist and environmental practices. Some of those interviewed were even explicit in their political statements about the connection, seeing themselves as helping to repair 'what Britain did' to its ex-colonies. But for this particular group of campaigners travelling interests, less than the colonial guilt, have always combined with what a Greenpeace campaigner defined as 'a scientific component, an adventure component, and a community component'. It seems as though, following this order, it is the adventure component which offers the attachment to science in practical form, so that the community dimension seems simply a consequence of the other two aspects of the experience. All this has a direct influence on how Brazil and the Amazon are perceived within this tendency. The major concern seems to be with the preservation of 'the biodiversity'. When the 'forest people' are considered – mainly indigenous people – it is in relation to the preservation of 'the forest', *sensu stricto*.

Although involved in the campaign against the mahogany trade from the Amazon for four and a half years, the Friends of the Earth campaigner only made her first journey to the Amazon shortly after the interview. The Greenpeace campaigner had made just a few weeks' tour by ship after over two years campaigning on the same topic. Although other campaigners from the same organizations, including WWF, had been on short trips to the Amazon for contacts with local NGOs, the point I am attempting to emphasize is that none of them had any particular direct links to the region. Expressing a 'sense of the globe', these campaigners were able to articulate forest issues within a framework of general economic and political structures. In other words, they were aware of global processes and politically articulate concerns about the impact of global economic practices upon particular forest areas. However, owing to their specific social and personal backgrounds, it appears that for this particular tendency forests are rather understood and conceptualized in an abstract and technical manner. The result of this is that specific historical, cultural and social realities are still predominantly subsumed under general models of forests. In this sense, the realities comprising the Amazon rainforest seem to be equivalent to other rainforest contexts on the globe, such as Malaysia and Indonesia.

It follows from this that the contact that 'trees' campaigners have with the 'local communities' or 'local people' – as they are generically referred to – is mainly a professional contact with 'local' NGOs which play a crucial role as suppliers of information to those organizations in the UK. As the FOE campaigner stated when I asked her if her organization was developing projects in Brazil:

> *We have close ties with FOE International in the Amazon. I mean, with the Amazon FOE group.* There is an organization known as GTA which is a network of about 200 Brazilian NGOs. They wrote to us and said: please, can you work on this situation that is really terrible, we need your help . . . But there are a lot of other groups like that. The socio-ambiental . . . something like that [the ISA – Instituto Socio-Ambiental, which is a national organization based in Brasília and São Paulo]. They are working to the same end, really. They are lobbying their government to try and halt the illegal logging of mahogany within the Indian reserves. They are taking on the human rights issue as well, the Indians being murdered and corrupted.

The priority of the agenda is indeed the mahogany trade. Although the Amazon is conceptualized in relation to its different ecosystems, there is a predominant focus on the *terre-firme* forests – the high canopy forests – which are immune to periodic flooding, by contrast with the Várzeas (floodplains). These *terre-firme forests* are more easily framed in messages for the general public in the UK. Moreover, reflecting the growing international concern with social justice and human rights issues, the defence of the 'local people' relates in this case to a particular population of the forest – the indigenous peoples in the Amazon. As the Greenpeace campaigner also states:

> *We've got an office, Greenpeace now in Rio, and we ally very very closely with them,* over coming up with, as a Greenpeace International Representative, to come up with international strategies to make sure the campaign is successful. We have a very good, strong group of NGOs in this country working on the mahogany issue. We will be calling for a moratorium on the trade of Brazilian mahogany on the basis that the information we have from Brazil, from the people on the ground: Indian communities and environmental groups there, as well as Greenpeace, that the majority of mahogany there is logged illegally in indigenous reserves.

Nevertheless, although she is speaking about indigenous peoples and human rights issues, these ideas seem to be used as a secondary argument to support the main campaign target. Sometimes it even appears as though the interviewee is formulating a strategic argument for legitimizing the campaign and mobilizing sympathy, as she further explicitly states when asked about Greenpeace strategies for the mahogany campaign:

> The first stage is to build awareness of the problem. We do that through mailings and also through media coverage by selling the story to newspapers, to the television and to the radio. To get the message out, so to educate people that there is a problem. Once people have been educated that there is a problem, and then *they can use* [the fact] *that the people are hungry or they want to do something, so you can use that 'people panel' so to speak, to have an effect on where it matches.*

At other times, 'local people' – as a generic abstraction – may also refer to Brazilian NGOs and grassroots groups, as when campaigners need to justify that their campaigns evolved from local people's demands. For this group of campaigners, the links promoted with the Amazon are mainly through the professional, well-established NGOs. The fact that the Amazon is a place imagined in technical terms – mainly through a forestry perspective – increases the distance between global NGOs and the local communities. English, along with technical concepts, is the language spoken by the environmental actors, hence there is a need for national intermediaries with the same skills. WWF, Greenpeace and FOE each have their own branches or offices in Brazil. As the Friends of the Earth campaigner told me:

> We can't afford to support a group that has just set up and wants to work in the rainforest. You know, we don't have the time, the money or the resources. We are working on ourselves. *But, if there is a group working on it as well as us, and can provide us information well, we will work with them. Like we work with Survival or Greenpeace. It's mutual, you see.*

Surely, there is a need to share principles and mutual goals in a co-operative project? However, the 'mutuality', which stands here for an equal relationship, is clearly defined and only understandable in relation to other transnational or UK-based organizations such as Greenpeace and Survival International. In fact, these NGOs set the standard or profile of what is expected from an organization when considering partnership working. However, questions arise when the other side is an organization from the Amazon. The global British-based NGOs' priority is to work with well-established professional NGOs, those with technical expertise, English and computer skills which enable them to carry out research and to provide information. Local Amazonian issues need to be translated, and the relationship with grassroots or local Amazonian communities needs to be transformed by the intermediary Brazilian professional NGOs in order to coincide with the global agenda of the transnational organizations. Furthermore, the majority of the intermediary NGOs are based in the south of Brazil and depend on the financial support of international groups. The questions that remain are: how far are the Brazilian NGOs able to address the needs of local agendas, at the same time as responding to the increasing demands for information from 'people on the ground' from the global NGOs? How much does the global agenda encompass or even erase the local demands? As the Greenpeace campaigner points out:

> *The strength of the campaign is that it is international. It has these powerful international links.* So, for example, in order for us to be running campaigns against mahogany in the UK, we need specific information from Brazil. Different, specific types of information. *We need it in a way that we can use it. There is no point in turning . . . producing something we cannot use, or it is inappropriate or whatever.* So, it is important that we sit down and talk about how to structure the international campaign; what we need from Brazil; to talk about the timing, or when things will happen or how we can help them.

The emphasis is in what 'we' – the UK campaigners – need. Of course this is justifiable in relation to the strategies for an international campaign. Governments, such as the Brazilian government, are vulnerable and responsive to international pressure. Moreover, there are certain limits within which global NGOs can act, and legitimize their actions, in order not to cross a nation state's sovereignty and sensitivity. The mahogany campaign tackles the international trade – the imports of mahogany from Brazil by the UK and the US – and this is how the UK campaigners can comfortably work for the preservation of the forest without interfering in 'domestic' problems. My question is to what extent the Brazilian NGOs face a situation of spending their time and resources providing information to meet the demands of the global NGOs, instead of focusing on local and regional priorities – such as the issue of timber itself (which is predominantly consumed in the internal market), along with the issues of land reform, agribusiness, mining, and all the consequent conflicts among different social and economic groups in the Amazon. In addition, are the well-established Brazilian NGOs considered 'global' after all? If they are, what is their role in setting up global campaigns and strategies? This remains open to further investigation.

Trees and people

Alongside those in the mahogany campaign and the mainstream environmental groups, there is a second tendency which tries to articulate the 'forest issues' in a broader, political way. The campaigners within this tendency represent an older generation – generally in their forties – with background training in a diverse range of fields related to the social sciences, literature, linguistics and the arts. In this sense they are a more diverse group. They are also dispersed in smaller organizations and networks such as Reforest the Earth, the Gaia Foundation and the World Rainforest Movement.

Rather than being technically orientated, their involvement with campaigning activity and their framing of the Amazon is related to ethical and political considerations developed since the 1970s. A common ground between them is that they have been influenced by publications such as the *Blueprint for Survival*, and often relate to the peace and women's movements as associated areas of interest and engagement. The majority have had some experience of living in 'Third-world' countries (in Africa and Latin America, but not necessarily Brazil) and have established personal and professional links to particular indigenous communities and local organizations. Concerned with the empowerment of 'local communities', their activities range from direct action and political lobbying, to supporting local projects and the networking of local groups – but they are mainly concerned with indigenous peoples, as are the 'trees' campaigners.

Hence, the experience of 'being there' for this group is connected more to the growth of their political awareness and their political militancy rather than exclusively related to their technical and professional qualifications. As a campaigner for the organization Reforest the Earth (female, forties) points out:

> *I became political, I think, just through going to Cameroon and finding out that there were things that were wrong which you had to stand up for and do something about.* Therefore I become more political as I get older. Because we have to take the power back from structures that are taking it away from us . . . I went out to live in Cameroon for three years. Because I wanted to do something called Voluntary Service Overseas. I was interested in – Third World development. You know, why we appeared to be rich and other countries appeared to be poor. *And I thought it would be interesting to go to a Third World country and find out about that . . . I discovered that there was – the fault, if you like, of the Western countries,*

> *my country, why these other countries were poor and found out about the unfair trading relationships. I was quite shocked to find that colonialism wasn't dead . . . And find there was still racism around and begin to understand the whole system whereby people were kept in their place.*

The personal experience of living in a 'Third-world' country, experiencing social and economic inequalities, helped to shape her political views and attitudes. The relation between different issues, such as gender, environment and peace and the links she establishes between them at international, national and local levels derive from that experience:

> What's become quite clear now is that, of course, the systems and structures around the world as a whole, it comes as a whole, have taken away the power from their own citizens as well. *So, it has become quite clear that poverty and the gap between rich and poor that we are finding in our country is connected to what I first noticed between the rich world and the poor world.* And it was that experience there and the beginning of the environmental movement here, with the first publication of the first ecologist magazine called *The Blueprint for Survival*, which made me aware of environmental issues.

Her travelling experience made her aware of systems of inequalities and power relations not only between her country and others, but also within her own country. Furthermore, her self-identity is formulated not only as an environmental campaigner, but through the experience of activism in the peace and women's movements:

> And I went to Greenham. It was a women's peace camp. It was outside of one of the nuclear bases, American nuclear bases. And it attracted many different women. It was made a women's peace camp because there were some rapes and things early on and they – and also the aggression against men from the military was much greater than against women. There were all sorts of reasons why it became women. That was the first place I got arrested ever. And got involved in blockades. And that empowered me, really, to go back and do it here rather than just stay there.

This Reforest the Earth campaigner is well known for her 'direct actions', which range from office occupations to what she calls 'ethical shop-lifting' in the case of mahogany, and have taken her to prison on several occasions. Politically she sees herself as coming from a Gandhian perspective of reacting in a non-violent way. For her, the central issue is to 'empower the disempowered at all levels'. She sees gender, peace, human rights and environmental issues as all interlinked, and linked internationally, nationally and locally. Similar to the 'trees' campaigners, she does not have any particular link with the Amazon region or its people. She has made a short visit to the Amazon in order to make contacts with local NGOs, governmental agencies and mahogany traders. She has never lived there, nor does she speak Portuguese. However, unlike the 'trees' people, her concerns are less technical and more concerned with global political and economic structures. Interacting or networking with the 'trees' tendency through the Forest Network-UK, her input into the mahogany campaign is to focus on how the international timber trade affects the life of the 'forest peoples':

> I have travelled a lot and seen and mixed with so many different people. And I see that the structures that we are fighting are all very similar – and if you don't think of the human beings as part of the ecosystem too, if you don't involve local people, if you don't involve social justice as well – if you are just thinking about conserving a tree or a frog, and you don't look at the whole thing, then you are not going to be able to save it.

Nevertheless, acting globally for a small NGO such as Reforest the Earth implies several restraints. The information coming from the Amazon is pretty much dependent on the level of networking with the other transnational NGOs, such as FOE, WWF and Greenpeace. Thus, the layers of intermediaries between local communities and this kind of organization increase enormously. Besides the nationally structured Brazilian NGOs, there is the intermediary of other global British-based groups. This difficulty is felt not only on the level of technical information, but also in the lack of communication owing to linguistic and cultural differences. Lacking knowledge of the local culture, history and language may imply a construction of generic categories, such as 'powerless' forest people, and so on, which are decontextualized and themselves displaced and emptied of their more precise meanings. It is also a form of distancing from local peculiarities that again, as with the previous tendency, allows for a political practice in global and structural terms. As such, 'the poor', 'the disempowered' and the 'local people' can be placed and replaced in different political struggles around global issues.

Another representative of this tendency is a former Survival International campaigner who now works in assisting the networking of local groups among themselves, and with global NGOs through the World Rainforest Movement. His personal, professional and political process of displacement took him to Africa and India before living for a few years in the Venezuelan Amazon as an anthropologist. His political awareness developed from his academic experience among the Yanomami people:

> I was trying to understand how people conceive their relationship with their environment and to what extent there is a parallel with their real relationship with their environment. It was an academic study for a doctorate . . . *and all the time I was doing these studies I became more and more concerned about the future of the people.* Because I could see what happened was that these people were being basically destroyed through

> their contact with Venezuelan society, the outside world. And we were aware of depredations by transnational corporations as well as by national businesses. All these people, and particularly the issue of land rights, the issue of health, because the Yanomami area had these terrible epidemics . . . and *I was a witness to the impact of these epidemics and realized that these people were in crisis, all sorts of crisis . . . I decided that I didn't want to be an academic. Because the academics didn't seem to really care. They were just studying this process without really intervening, and I felt that we had the moral obligation to intervene in this process which was obviously injust, and certainly deadly. And so I converted into a . . . I keep becoming a human rights activist.* (World Rainforest Movement, male in his forties)

'Being there' for this campaigner brought him ethical and moral dilemmas which were rationalized in a political manner. Despite calling himself a 'human rights activist', he also classifies himself as a 'green':

> I was familiar with different cultures, with different ways of seeing things, with different languages, and I was also familiar with the fact that, you know, *the English were not necessarily the best of everything at all. You know, there are other societies, so there are different answers to the same questions* . . . and in many ways, when I went back to the Amazon in the 1970s, I was looking for alternative societies to explain what went wrong with ours and because even from the age of 16 I felt very alienated from the western civilizations. *I was sure it was flawed and wrong, and I've always been a green in the sense that I've always thought that . . . basically, capitalism is very materialistic and unroading in the long term. And it denies many other human values. And so, I was particularly interested in other human societies which maybe were driven by different forces.*

It is possible to identify parallels in this statement with the other campaigner with regard to links between their familiarity with different cultures and their critical view of their own society, and the power relations operating in the global realm. In this particular case, his political identity as 'a green' who belonged to a generation of protesters and of readers of *Blueprint for Survival* in the 1970s (the same literature mentioned by the other campaigner) is also matched by his readings of Amazonian travellers, such as Alexander von Humboldt. 'Being there' for this campaigner drove him closer, in personal terms, to indigenous peoples and the problems faced with the 'outside world'. His links are more strictly constructed with indigenous people and their relationship with their environment – an exotic culture compared with his own – perhaps perceived as a more 'authentic' one in contrast to the capitalist societies which he classifies as 'wrong'. His primary focus seems to be on indigenous people and the impact they suffer from changes in their environment.

Similar to the statements of the Reforest the Earth campaigner, reflecting on the need to link up people and the environment, he expresses a criticism of the conservationist groups which he sees as particularly concerned with rainforest destruction without considering the point of view of social justice and the local people. However he perceives this as a past approach and with optimism points out that there has been a change in this kind of approach with the incorporation of such issues by the environmental movement at large.

When asked about his activities as a campaigner and his relationship with local people he explains:

> *Well, it's all kinds of different things, because it depends on what they need, or what they want, or what they ask for.* So, it may be that they want to gain access to the International Tropical Timber Organization, because their lands have been logged under projects supported by the ITTO. So, we would facilitate them to go to international meetings to lobby for their rights . . . *So, we raise money so they can represent their own interests directly in negotiations with international institutions . . . We work to promote their own dialogue with aid agencies that are creating national policies.* So, institutions like the World Bank maybe want them to develop a natural resource policy for a country . . . and the local communities may think that they should at least be consulted [laughing]. And so, we try to help them get information about that process . . . the majority of the work is on providing technical assistance in policy debates.

In contrast to the 'trees' tendency, the emphasis here is in what 'they' – indigenous communities – need. However, it is possible to say that there is a 'division of labour' in the Forest Network-UK field. The umbrella organization he works for can encompass the other NGOs, but one of the differences between this approach and that of individual groups is the two-way flow of information. Hence, in contrast to the one-way flow of information from local to global which characterizes most NGO relationships, information about global processes which affect local communities is given from global NGOs to local groups rather than simply the other way around. As a network of NGOs, information in this case circulates more local-globally, global-locally, local-locally than in other cases. Nevertheless, there is also the possibility of personal advocacy in specific situations and issues regarding international laws, or the activities of transnational corporations affecting indigenous peoples. His justification for his direct lobbying and advocacy is given in reference to the political context and criticism from nation states in the 'South':

> *Unfortunately, it is the case that, you know, many of these governments tend to pay more attention to a Northern environmentalist than to their own people.* Even though one's saying exactly the same thing, they take more

> notice . . . Some of these countries still require a lot of aid, and aid agencies have now got a concern about the environment. And some of them have got a concern about indigenous rights. *The governments feel that the Northern campaigning organizations can affect how much aid they get if they don't show willingness. So, there is a conditionality in aid which can in fact support claims of local communities* . . . so, they use the international campaigning organizations to create what they call 'conditionality' to impose conditions on the government which won't listen to their own people. So, because they won't bother to listen to their own people it's gone this whole laborious route round. *So, what the governments have done is actually, by not listening to their own people, they've recreated their own colonialism. Because they now have been told what to do by the North.* Because they won't listen to their own people in their own countries.

His remarks illustrate the relevance of support from 'Northern' NGOs in the claims of local communities that they are marginalized by 'Southern' governments driven by a particular hegemonic 'developmental' agenda, which is itself also set up by global financial agencies. Moreover, it articulates a clear response to the allegations and accusations of 'eco-colonialism' by 'Southern' governments and local economic elites. As I have already mentioned, this was a recurrent topic during my interviews and a frame of reference for the discourses of the campaigners interviewed. By rationalizing and legitimizing their political engagements in global terms, whether through references to the field of international trade and other global power structures, or by stating ethical and moral responsibilities and solidarity with 'indigenous peoples' and 'local communities', they are often aware of the Brazilian sensitivity to the issue of transnational campaigning for the Amazon.

Across the range of environmental NGOs, active within the Forest Network-UK, this 'trees and people' tendency is still in the minority, most campaigners tending to follow the same agenda that is set up by the major environmental NGOs. Networking with the major British-based groups provides those with fewer resources the means to acquire information on 'local contexts', as well as to acquire the technical information they otherwise lack. On the other hand, the smaller groups have contributed valuable political input and skills to the campaigns run by the larger organizations. Nevertheless, lack of resources from smaller NGOs as well as prevailing interest from the major NGOs in the 'trees' tendency led to the end of the Forest Network-UK in 1997, after a few years of productive exchange between different groups engaged in the mahogany campaign.

People

Campaigners within the 'people' tendency have a political agenda related to the advocacy of human rights, social justice or 'social development' in 'Third-world' countries. They are active within the Brazil Network and are not particularly involved in forest campaigns, although they might express concerns about the Amazon forest as a place in which people gain livelihoods. They work for organizations such as Oxfam, Cafod, Christian Aid, Amnesty International and Survival International – this latter group keeping an interface with campaigners of the tendencies discussed above.

Generally speaking, they suggest some similarities to campaigners of the 'trees and people' tendency, particularly in terms of their professional qualifications and age composition. However, in contrast to those, they have a background history of political involvement with Latin America and Brazil. They have usually lived in Brazil, speak the language and have acquired significant knowledge and understanding of regional and domestic contexts, as well as considering their place in global dynamics.

In contrast to the 'trees' campaigners with expertise in forest issues, these campaigners can be defined as being 'Latin Americanists' or 'Brazilianists'. However, they believe they maintain a dialogue with the 'environmental' campaigners and, besides certain criticisms of the others' possible lack of 'knowledge' and acts of 'political compromise' to the Brazilian context, they believe they have politically influenced many of the changes within the field of environmental discourses. The views of this group can be very close to those of the previous campaigners in the defence of indigenous peoples' rights, but they also encompass the other forest social groups and movements. A common and underlying campaign for this tendency is the campaign for 'land rights'.

Whether an indigenous rights campaigner, with a discourse very close to that of the World Rainforest Movement campaigner, or a campaigner for the peasant and the landless people in the Amazon (i.e., the local communities such as the rubber tappers and the riverine), a key commonality among the campaigners in this tendency is that they have personal ties or have had personal and professional experience in Brazil or in the Amazon. Alongside the links with local NGOs, they have also established links with local communities, as well as with grassroots and other social and political movements.

This is the case with a campaigner from Survival International, an organization from this 'people' tendency that has close interaction with environmental organizations. However, the closest personal and professional ties are to those campaigning for social and economic rights in the Amazon:[19]

> I lived in Roraima [a state in the Amazon region], precisely where there are many indigenous problems, for fourteen months. I worked in an environmental project which was run by Impa, in Manaus, and the Royal Geographical Society in London. So, it was a great experience . . . I heard about the project and I wanted to go back to Latin America. Because I did some academic research in Peru, with the Quechua people, and I spoke a little bit of Quechua. *And also, my parents used to live in Venezuela. So, I always felt myself to have very close ties to Latin America . . . and then, when I was a child I spent five years in South Africa,*

> *and then I think that it comes from there, my awareness of racism and the lack of basic human rights during apartheid. I believe it raised in me the will to be more involved in the minorities' struggles.* (Survival International, female, thirties)

The displacement experience made her politically aware of the problems faced by minorities, and while living in Latin America the minority in question was the indigenous population. As with others in this group, although also going into 'the field' as a researcher, it is in the areas of humanities or social sciences that they find their interests, and not in the field of forestry. This is also the case with a former Oxfam and Rainforest Foundation campaigner. Belonging to what she defines as 'an older generation' of activists, in comparison with that she calls 'the greens', her involvement with the Amazon came as a result of her political experience in South America, as a socialist supporter of Salvador Allende in Chile in the early 1970s:

> *In '72 I went to Chile. It was my student's adventure.* I wanted to know Latin America. I wanted to know Allende, the country of Salvador Allende. Then I was studying. I was going to do a PhD in Chile. But I studied very little. I spent more than one year, one and a half years in Chile, at that time of Allende. I was there during the state coup in Chile, right? September '73, right? Then I met many Brazilians, Latin Americans from many countries who were in exile there . . . *It* [socialism] *is a kind of romanticism again* [laughs]. Then I decided to come back. I had few options because I had finished my studies with the state coup . . . I came back after the coup and went to work for the international office of Amnesty International. I will never forget, because my first work, one of my first works at Amnesty was to prepare the campaign of the ten years of coup in Brazil, right? . . . Researching the number of political prisoners, tortured, disappeared in Brazil. (female in her forties)

Thus far, from the beginning her involvement with Brazil came as a direct result of her socialist views and interests. The political experience in Chile referred to as a 'student's adventure' places her in the context of the European–Latin American leftist engagement and political solidarity movement of the 1970s, by contrast to the 'environmental' flow of late 1980s and 1990s. Hence, it seems as though the contact with 'the exotic', which for the 'trees' campaigners has a political meaning related to the 'exoticism' of biodiversity, or even the indigenous populations – a utopia of 'nature' in symbolic terms – for this group assumes the meaning of an 'exotic' political regime or a symbolic social utopia. The 'socialist' background is a topic treated with irony, perhaps reflecting the political and ideological fragmentation of the 1990s. The student's romantic adventure in Chile was thus finished with the state coup in that country. Back in England, the student became a professional campaigner for NGOs, whose interests gradually shifted from the advocacy of political human rights to a broader, more economic and social perspective:

> Then I left [Amnesty] in 1990. I needed to move. My considerations about the meaning of human rights changed. Amnesty only works about political rights, right? Political and social. And now the issue of economic rights is not worked by Amnesty International. And I was interested in that. I wanted to get into another field, to change and to broaden my knowledge . . . I was a very active trade unionist. Especially in the 70s and 80s in this country, I was a campaigner for trade unionists and workers who were in prison and persecuted. *I suppose that was the kind of way I moved. Because the main problem for the trade unionists, workers, peasants who get persecuted is that they are fighting for a better life. And a better life – as well as being more political freedom is also more economically based – better, better life, you know? A bit more money* [laughing], *more wages, more land . . .*

The topic of romanticism emerges again when she refers to her connections to the Amazon forest. Her first involvement with the Amazon was while working on the development of health and agricultural projects among two indigenous groups in Peru:

> So, my first contact with indigenous forest peoples was with those two peoples. And I grew to love them and respect them enormously. *To love their country, their land, the forest, the trees, the rivers, the insects. The bad thing, the heat, all the bad things that most Europeans hate, the insects – all these things . . . And so, there is a level of romanticism. I'm still a totally incurable romantic.* I'm not – I'm not afraid of saying that – but I'm also – working in this area, so I have to try to balance my romanticism with a degree of rigorous objectivity as well.

Her self-image as a romantic person is ironically contrasted to those who she believes are considered romantic by many Brazilians: 'most European environmentalists'. The reason why she emphasizes her personal and direct experience of the 'bad things of the forest', particularly the insects, is because she knows that there is a general view in Brazil that Europeans concerned with the preservation of forests are romantic and, in many cases, form their views without ever having visited a forest. She tells me this because she thinks it is something we both acknowledge, although she is not sure about my position as a Brazilian. There is also an implicit criticism, or an opposition to the technical environmentalists. She believes that her practical experience legitimizes her involvement with Amazon people: she has 'been there' through her work; therefore, despite her 'romanticism', she has 'objective' knowledge and experience of what she campaigns for.

Objectivity and professionalism are highlighted as those attributes currently valued for professional campaigners amidst the proliferation of NGOs.

This can be further observed through the statements of the Survival International campaigner:

> We think that to achieve positive changes in the world, you have to influence, at the end of the day, the governments, because they are the ones making policies. And the best way to do it is through public pressure. But we do it in a peaceful way and the Brazilians know that. We are respected for our work because *we work in a serious way. We are respected because we are always trying to be in the field. I go there and I speak to a lot of people. It is very important to research and to publish after carrying out a rigorous investigation of the facts.* We have arguments with governments as well as institutions such as the World Bank. We write letters to the different responsible bodies and *we are very respected for the accuracy of our information.* It is very important to have direct contact to the area, to speak directly to the indigenous groups and also to the Brazilian organizations working at the grassroots level.

There is an implicit reference here to those various environmental organizations which appeared in the 1990s following the Rio Earth Summit Conference of 1992. However, the debate between campaigners of different organizations can be traced back to before the Conference. Belonging to an 'older generation' of political campaigners in the UK with close ties to Latin America, a former Oxfam campaigner tells me of her experience in political debates in the mid-1980s with what she classifies as the 'new generation of political activists in Latin America – the 'greens':

> In the years of '86, '87, '88, we were very few who really used to speak about the social issues in the Amazon. The images from the media here were very frustrating and negative. There was only the forest, the green, the animals, the trees, the burning . . . *I became very critical and angry about those documentaries. The Indians were not people. They were exotic in the exotic forest: 'we must protect the Indians, the animals, the trees, the river, etc'. They were all equal. It was horrible.* Up till today, the media does not acknowledge that in the Amazon there are cities, big cities which emerged in this period in the 1980s. But to speak about cities to the public here in the context of the Amazon, ecology, social development – forget it.

The campaigns concerned with 'Amazonian people' not only are related to the indigenous populations, but encompass other social groups such as peasants and rubber tappers, tackling issues of land rights and fair trading. The Brazil Network, in which the 'people' campaigners act, also hosts a range of individuals such as journalists, film-makers and academics involved with Brazil. As expressed above, there is a certain cleavage between the campaigners within this 'people' tendency, and those concerned with biodiversity issues in the Amazon. There is even a certain dispute in terms of political legitimacy between those who perceive themselves as 'Brazilian experts' and long-term militants, and the 'new' political actors represented by the environmentalists. Environmentalists, and events such as the Earth Summit, are resented for having 'stolen people's agenda'. Nevertheless, if this group was very critical of environmentalists in the mid-1980s for their lack of sensitivity to social, economic and political issues related to the Amazon, environmentalists for their part might perceive this group as acting as 'charities' on a very localized, temporary and fragmentary basis – as aid agencies rather than as campaigners for transformations on a broader scale. These differences apart, most campaigners for the Amazon within what I have called the three tendencies in the UK do express a vision about global processes and the related effects upon 'local people'. The debate and disputes among them have contributed to a mutual fostering of new perspectives, dynamics and actions in global as well as in local policies towards the Amazon and its people.

Conclusions

I have classified British-based campaigners for the Amazon into three major tendencies: those who present discourses orientated to '*trees*', '*trees and people*' and '*people*'. They have been identified according to their focus on environment/biodiversity on the one hand, and social justice/development on the other, as well as in relation to their personal experiences of displacement, their relationships with 'the local' and their justifications for their actions. Hence, beyond the obvious differences between environmental and developmental NGOs, the idea has been to focus on campaigners who are actually the carriers of NGO ethos, and therefore important actors in shaping the debate about the Amazon.

In short, all three tendencies identified here do articulate 'trees and people', that is, issues of biodiversity, social justice, human rights and social development. These aspects are said to have been reconciled since the late 1980s, particularly with the Brundtland Report and its concept of sustainable development, which was further consolidated through the Earth Summit in 1992.

Since the time the UN conference was held (significantly enough, in Brazil) an official environmental discourse has been firmly established. Nevertheless, this discourse seems to inscribe society and development within a totalizing and evolutionist concept of economic growth, whereas 'nature' appears to be merely as a variable to be 'managed'. These global tendencies have imposed themselves upon peoples, societies and environments. The 'development' ideology seems to be reinvigorated by the concept of 'sustainable development', which also claims respect for biodiversity as well as cultural diversity. In order to legitimize this official discourse, NGOs – which were longstanding holders of a counter discourse – were invited into 'participation' and 'partnership'. Within this general framework and trend, the mainstream NGOs seem to have thus accommodated their discourses into such an encompassing and institutionalized formation.

The idea of 'sustainable development' – vaguely meaning considering people's needs beyond conservationist practices – has its general principles agreed upon. But it meets concrete obstacles when faced with practical initiatives and implementations. As far as forests are concerned, particularly the Amazon, different approaches towards sustainable development make it difficult to uphold the simplified and common classification of NGOs into categories such as, for example, conservationist and environmental organizations. Working with 'local people' or 'local communities' has become a rather popular claim. Different positions in the British-Amazonian field are influenced by campaigners' social and personal trajectories, which lead them to participate in discourses that emphasize one aspect of the field or another. I have specified regularities and discursive patterns through discussing the campaigners' experiences as travellers, their professional training, generation, political awareness and involvement with the Amazon and Brazil. If the differences reveal a diverse and dynamic campaigning field, they also unveil some problems of communication and exchange to be overcome by campaigners among themselves in the UK, as well as with their Brazilian counterparts.

The two tendencies grouped in the Forest Network-UK – 'trees' and 'trees and people' – appeared primarily concerned about the impact of global economic practices upon particular forest areas, and they have been very effective in their actions. However, it seems that the technical expertise orientation of campaigners under the 'trees' tendency – a necessary element of the NGO work in their counter-arguments to policy makers and business interests – along with a global perspective mainly focused on biodiversity, and a sense of cultural distance from particular social and historical contexts in the Amazon, are the main problems to overcome. The Amazon is cut off from its historical, local and regional contexts and often projected into the global arena as a mere ecosystem, under the influence of global economic and political forces. Longstanding historical patterns of policies in the region, as well as different social systems in the Amazon, become invisible or subsumed in relation to technical abstractions and definitions under global patterns. Hence, although incorporating the defence of 'local people' (meaning mainly indigenous people rather than Brazilians of immigrant origin) in their discourse, it still remains as a secondary topic in their considerations. 'Partnership' is established with Brazilian NGOs that can respond to the appropriate requirements, namely, technical expertise and computer and linguistic skills. Owing to the demand for information coming from NGOs in Britain, and the level of economic dependency of Brazilian NGOs upon their transnational partners, it is difficult to tell the scope and the role of Brazilian NGOs in effectively setting up the dynamics of the transnational agenda.

'Trees and people' campaigners are an older generation, less technical, more diverse, and are spread out across various smaller organizations and networks in comparison with the 'trees' campaigners. They tend to be more politically committed to supporting local perspectives and alternatives. However, the situation of intermediation between 'local' and 'global' perspectives concerning these campaigners tends to be more pronounced. Most have fewer resources and depend on the information the larger groups can provide. Therefore they have in general a limited impact on setting the overall agenda in the UK, although they contribute valuable elements to the established campaigns. Sometimes, as in the case of long-established campaigners, they can influence the direction of the overall debate among campaigners towards positions more in tune with NGOs' original principles, particularly when campaigners and campaigns seem to get too subsumed in negotiations with the private sector and governments.

The 'people'-orientated tendency revealed a political agenda related to the advocacy of human rights and social justice. They are grouped in the Brazil Network and are not necessarily involved with environmental campaigns. Generally speaking, they have a background of training in the humanities and also a history of political socialist or feminist involvement with Brazil and Latin America, rather than a professional training in forestry, or environmental activism. Concerns with inequality, social justice and economic distribution in Latin America may appear at times to drive them closer to a discourse associated with an ideological perspective centred on 'development', though they are usually opposed to developmental projects which have long favoured local elites and transnational capital. Establishing close personal and political ties to the 'local people' and acquiring some anthropological sensibility, they appear as 'Latin Americanists' or 'Brazilianists'. Therefore, they can present themselves as the only 'authorized voice' about that region. The ethical and political concern with 'people' – through attention to the inequalities between the 'Northern' and 'Southern' hemisphere – requires a practical approach to campaigning which makes them be perceived by 'trees' campaigners as less concerned with biodiversity and a 'global' environmental perspective that would more firmly question 'development' as a remedy for poverty. Their agenda is, in fact, more concerned with the support for the organization of grassroots and social movements in Brazil. A possible challenge for them is to establish a dialogue with the other tendencies in order to incorporate environmental concerns more generally into their own campaigns for the people of the forest, and also to push the social component into the environmental debate.

Finally, it is relevant to notice that environmentalism and ethnicity are intertwined as far as the Amazon is constructed in the UK, and in the transnational arena at large. The advance of the western economy over the rainforest and indigenous peoples' territories has led to the representation of indigenous peoples' interests beyond their traditional local sphere. This coincides with environmentalists' claims and concerns with biodiversity and the global environment.

In the late 1980s the burning of the forest was the commonest image of the Amazon spread across the media and the campaigns of environmental groups. In the 1990s, it is logging activity in the Amazon which preoccupies the agendas of most British environmental NGOs. The campaign against the mahogany trade heightened the need for special working links between NGOs based in Brazil and in Britain, since the UK is regarded as one of the major importers of that particular timber. Undoubtedly, Non-Governmental Organizations have become one of the most important political forces in recent years, challenging traditional and nation state based forms of political organization. Environmental organizations are among these and their technical expertise is a necessary tool in terms of forest management and policies. Nevertheless, a crucial future challenge for environmental groups will be to escape from the asepsis of the technical-scientific distance from local cultural contexts, which seems to have driven them currently into the trap of the liberal market-orientated agenda.

Notes

1 J. Clifford, *The Predicament of Culture: Twentieth-century Ethnography, Literature, and Art* (Cambridge, MA, 1994 [1988]), 263.

2 However, this is not a case of the 'anthropologised turning into anthropologist' – the reprisal of the Indian, the *caboclo* or the riverine – since I am not a Brazilian from the Amazon. My particular standpoint stems from the fact that Brazil is a country which has historically been an object of European and North American studies, whereas England, in particular, has always epitomized the European Other for Brazil. The peculiarity of my social and historical location is certainly relevant to the present discussion, and I hope this will become clear in the following pages. See further, A. Zhouri (1998) 'Trees and people: an anthropology of British campaigners for the Amazon rainforest', unpublished PhD thesis (University of Essex). Earlier versions of this chapter were presented at the IX International Oral History Conference, Göteborg, Sweden, 13–16 June 1996, and the III International Conference of the Brazilian Studies Association (BRASA), Cambridge, UK, 7–10 September 1996.

3 For different approaches see, for instance, K. Eder, *The Social Construction of Nature* (London and New Delhi, 1996); C. Muller, B. McClurg and A. Morris (eds), *Frontiers in Social Movements Theory* (Yale, 1992); M. Edwards and D. Hulme, *Non-Governmental Organisations: Performance and Accountability – Beyond the Magic Bullet* (London, 1995); M. Eyerman and A. Jamison, *Social Movements: A Cognitive Approach* (London, 1991); H. Johnston and B. Klandersman, *Social Movements and Culture* (London, 1995); A. Mellucci, *Nomads of the Present* (Philadelphia, PA, 1989), among others. For a critical discussion and an anthropological perspective see K. Milton, *Environmentalism and Cultural Theory: Exploring the Role of Anthropology in Environmental Discourses* (London, 1996).

4 I am using the word campaigner to state the professional character of those working for NGOs, in opposition to activists as those lay or non-professional political subjects. NGO is also a problematic concept, as it has been discussed by many authors (see, for instance, Edwards and Hulme, *Non-Governmental Organisations*). While acknowledging the distinctions between Non-Governmental Organizations, volunteer organizations or grassroots organizations, I will not discuss them within the limits of this chapter. For the current purpose, I will use the term NGO as it is employed by the campaigners themselves as part of their cultural and political identity as a collective actor, regardless of their working for highly structured and hierarchical organizations such as Friends of the Earth, Greenpeace, WWF, or small volunteer organizations such as Reforest the Earth. For an analysis of collective identity, see A. Melucci, 'The Process of Collective Identity', in Johnston and Klandersman, *Social Movements*.

5 B. Anderson, *Imagined Communities: Reflections on the Origins and Spread of Nationalism* (London, 1991 [1983]).

6 See for example, R. Robertson, *Globalization, Social Theory and Global Culture* (London, 1992); J. Friedman, *Cultural Identity and Global Process* (London, 1994); M. Featherstone, *Global Culture: Nationalism, Globalism and Modernity* (London, 1991); D. Harvey, *The Post-Modern Condition* (Oxford, 1989); A. Gupta and J. Ferguson, 'Beyond Culture: Space, Identity and the Politics of Difference', *Cultural Anthropology*, 7(1) (1992); A. Giddens, *Beyond Left and Right* (London, 1995).

7 U. Beck, *Risk Society: Towards a New Modernity* (London, 1992); Milton, *Environmentalism*; S. Lash, B. Szerszynski and B. Wynne, *Risk, Environment and Modernity: Towards a New Ecology* (London, 1996).

8 Giddens, *Beyond Left and Right* (London, 4).

9 Ibid. 7.

10 The in-depth interviews followed a general guide which was divided into two main blocks of questions. The first block was concerned with the interviewees' personal backgrounds, their involvement with environmental issues, NGOs and the Amazon. The second block was a thematic one concerned with the NGOs in terms of their specific campaigns, their relationships with the other NGOs in the UK and their counterparts in Brazil, and so on.

11 For the concepts of field and discourse that inspire my approach, see P. Bourdieu, *Outline of a Theory of Practice* (Cambridge, 1993 [1977]); and further, M. Foucault, *The Archaeology of Knowledge* (London, 1972).

12 See further, W. Sachs (ed.), *The Development Dictionary: A Guide to Knowledge as Power* (London, 1992); and J. Kirby *et al.*, *The Earthscan Reader in Sustainable Development* (London, 1995).

13 This has occurred especially since the MDB (Multilateral Development Bank) campaign in the early 1980s, and was also reinforced later on, during the UNCED-92 process, through the NGOs' closer relationship with social agencies already dealing with such matters. For an analysis of the MDB campaign, see A. Kolk, *Forests in International Environmental Politics: International Organisations, NGOs and the Brazilian Amazon* (Utrecht, 1996).

14 On 'discourses' and 'displacements', see Clifford, *The Predicament*. For a perspective on imperialism and the history of environmentalism, see A. Grove, *Green Imperialism: Colonial Expansion, Tropical Island Edens and the Origins of Environmentalism, 1600–1860* (Cambridge, 1995); and also D. Arnold, *The Problem of Nature: Environmentalism, Culture and European Expansion* (Oxford, 1996).

15 For a discussion on the relation between territory, state, economy and political representation in the context of the Amazon, see contributions to M. A. D'Incao and I. M. da Silveira (eds), *A Amazônia e a Crise da Modernidade* (Belém, 1994).

16 On the issue of 'local knowledge', see C. Geertz, *Local Knowledge* (New York, 1983).

17 All the emphasis in this and following quotations is mine.

18 In *Heart of Darkness*, Conrad writes:

> Now when I was a little chap I had a passion for maps. I would look for hours at South America, or Africa, or Australia, and lose myself in all the glories of exploration. At that time there were many *blank spaces on the earth*, and when I saw one that looked particularly inviting on a map (but they all look that) I would put my finger on it and say, When I grow up *I will go there.* The North Pole was one of these places, I remember. Well, *I haven't been there yet*, and I shall not try now. The glamour's off.
>
> J. Conrad, *Heart of Darkness* (London, 1995 [1902]), 21–2. The emphases are mine.

19 This interview was conducted in Portuguese and later translated.

Select bibliography

Anderson, B., *Imagined Communities: Reflections on the Origins and Spread of Nationalism* (London, 1991 [1983]).

Arnold, D., *The Problem of Nature: Environmentalism, Culture and European Expansion* (Oxford, 1996).

Beck, U., *Risk Society: Towards a New Modernity* (London, 1992).

Bourdieu, P., *Outline of a Theory of Practice* (Cambridge, 1993 [1977]).

Clifford, J., *The Predicament of Culture: Twentieth-Century Ethnography, Literature, and Art* (Cambridge, MA, 1994 [1988]).

Conrad, J., *Heart of Darkness* (London, 1995 [1902]).

D'Incao, M. A. and da Silveira, I. M. (eds), *A Amazônia e a Crise da Modernidade* (Belém, 1994).

Eder, K., *The Social Construction of Nature* (London and New Delhi, 1996).

Edwards, M. and Hulme, D., *Non-Governmental Organisations: Performance and Accountability – Beyond the Magic Bullet* (London, 1995).

Eyerman, R. and Jamison, A., *Social Movements: A Cognitive Approach* (London, 1991).

Featherstone, M., *Global Culture: Nationalism, Globalism and Modernity* (London, 1991).

Foucault, M., *The Archaeology of Knowledge* (London, 1972).

Friedman, J., *Cultural Identity and Global Process* (London, 1994).

Geertz, C., *Local Knowledge* (New York, 1983).

Giddens, A., *Beyond Left and Right* (London, 1995).

Grove, A., *Green Imperialism: Colonial Expansion, Tropical Island Edens and the Origins of Environmentalism, 1600–1860* (Cambridge, 1995).

Gupta, A. and Ferguson, J., 'Beyond Culture: Space, Identity and the Politics of Difference', *Cultural Anthropology*, 7(1) (1992), 6–23.

Harvey, D., *The Post-Modern Condition* (Oxford, 1989).

Johnston, H. and Klandersman, B., *Social Movements and Culture* (London, 1995).

Kirby, J., O'Keele, P. and Timberlake, L. U., *The Earthscan Reader in Sustainable Development* (London, 1995).

Kolk, A., *Forests in International Environmental Politics: International organisations, NGOs and the Brazilian Amazon* (Utrecht, 1996).

Lash, S., Szerszynski, B. and Wynne, B., *Risk, Environment and Modernity: Towards a New Ecology* (London, 1996).

Melucci, A., *Nomads of the Present* (Philadelphia, PA, 1989).

—— 'The Process of Collective Identity', in Johnston, H. and Klandersman, B., *Social Movements and Culture* (London, 1995), 41–63.

Milton, K., *Environmentalism and Cultural Theory: Exploring the Role of Anthropology in Environmental Discourses* (London, 1996).

Muller, C., McClurg, B. and Morris, A., *Frontiers in Social Movements Theory* (Yale, 1992).

Robertson, R., *Globalization, Social Theory and Global Culture* (London, 1992).

Sachs, W. (ed.), *The Development Dictionary. A Guide to Knowledge as Power* (London, 1992).

Zhouri, A., 'Trees and people: an anthropology of British campaigners for the Amazon rainforest', unpublished PhD thesis (University of Essex, 1988).

Reviews

'Not otherwise touchable somehow': Ecocriticism and literature

Jeff Wallace

In this short piece I want to identify a set of key texts and issues in the continuing development of what has become known as 'ecocriticism'. Any ecologically motivated perspective will be characterized by a certain inclusivity, and the proferred definition of ecocriticism by one recent practitioner – 'the new environmentalist cultural criticism' – indicates the already established breadth of some writing in this tradition.[1] However, for the purposes of economy, I will focus on ecocriticism as a movement within literary criticism, basing my discussion on British and American writing, and paying particular regard to two texts: Jonathan Bate's *Romantic Ecology: Wordsworth and the Environmental Tradition* (1991), and *Writing the Environment: Ecocriticism and Literature* (1998), edited by Richard Kerridge and Neil Sammels.

What do literary texts look like, when read from the perspective of an ecological commitment which takes it as axiomatic that we live in a time of profound environmental crisis? In asking this question, ecocriticism takes its place alongside feminism, Marxism, poststructuralism and postcolonial studies in insisting that literary criticism is always an ideologically *interested* activity, and that active interpretative commitment can enrich and transform our understanding of literary texts. Unlike its companions, however, we find in ecocriticism a movement not yet secure in the theoretical groundings of its relationship with literature, nor yet enjoying mainstream academic status. If the interaction between human kind and the natural world is acknowledged as one of the general structuring concerns of much imaginative literature, it would seem that the subject matter of ecocriticism is already securely established. Conversely, at least one contemporary ecocritic holds '(im)possibility' to be the paradoxically enabling and defining condition of this embryonic critical discourse.

To the extent that ecocriticism is a consciously interventionist movement, the seminal British text is undoubtedly Jonathan Bate's *Romantic Ecology*.

The book provoked lively debate, and much subsequent British ecocriticism has taken the form of essays which respond or refer to Bate in various ways. A very recent addition, Kerridge and Sammels' edited collection *Writing the Environment*, which itself contains an essay by Bate on 'Poetry and Biodiversity', shows signs that the influence of the earlier text may now be settling into the critical background. In America, a collection of single monographs form the backbone: of these, Karl Kroeber's *Ecological Literary Criticism: Romantic Imagining and the Biology of Mind* (1994) is close to Bate in its polemical drive and political position, while Lawrence Buell's *The Environmental Imagination: Thoreau, Nature Writing, and the Formation of American Culture* (1995) has already been highly influential. From the same year, Patrick D. Murphy's *Literature, Nature, and Other: Ecofeminist Critiques* is equally significant, and theoretically distinctive. One effect of these 1990s works, joined in the self-conscious project of establishing a movement of literary ecocriticism, is to invite speculation on the prehistory of ecocriticism – to trace, in other words, the ecocritical before ecocriticism. Raymond Williams' *The Country and the City* (1973) would be a strong contender in any such tradition; my own selection would include John Barrell's superb early study of Clare, *The Idea of Landscape and the Sense of Place: An Approach to the Poetry of John Clare* (1972). However, as each of the foregoing works belongs to a critical school with which Bate's seminal work is explicitly at odds, it will now be useful to outline Bate's position in more detail, and use this as a platform from which to explore some of the issues which inform literary ecocritical debate.

Written in the wake of the Soviet Union's demise, Bate's book takes a trenchantly adversarial stance towards that mode of criticism, New Historicism, which it takes to be an expression of an outmoded Cold War politics. Bate proposes that New Historicist readings of Wordsworth, such as those of Jerome McGann, Marjorie Levinson and Alan Liu, distort the poet's political significance by judging it through a paradigm of 'romantic ideology'. According to this paradigm, the romantic imagination works to transcend history and politics through the idealization of 'nature' and subjective experience. The Wordsworth this produces fits, for Bate, rather neatly into the traditional young radical/older conservative model of the poet; his poetry comes to be scanned for giveaway instances of the transition into false consciousness. Bate too seeks a thoroughly politicized Wordsworth, but, by contrast, through the reclamation of the 'Poet of Nature', as named by Shelley and known by the Victorians. Wordsworth's pastoralism, his concern for harmonious relationship with and understanding of nature, goes 'deeper than . . . the political model we have become used to thinking with', and makes his recuperation particularly apposite at a time when 'many ex-Marxists' in Europe are making 'the move' – somewhat audaciously sketched, perhaps – 'from red to green'.[2] Readings of *The Excursion*, with a notable section on 'The Ruined Cottage', seek to establish how far for Wordsworth pastoral poetry and egalitarian politics go hand in hand; the chapter 'The Moral of the

Landscape' shows how Ruskin learnt his ecological vision from Wordsworth, and 'The Naming of Places' advances an argument about the intertwining of poetic language and geographical location which prefigures its more complex treatment in the later essay on modern poetry (Edward Thomas, Basil Bunting, Seamus Heaney and Les Murray) and Heideggerian indwelling.

For Bate, then, to go back to Wordsworth is simultaneously to move forward into a politics which transcends the old oppositional politics of left and right; we, or at least the New Historicists among us, have forgotten what Wordsworth was really about, because of our insistence that writers should pursue a critical appraisal of socio-economic reality. One recalls Graham Holderness' claim in 1982 that D. H. Lawrence in *The Rainbow* made 'no attempt to describe a society that ever existed', his mythic depiction of Marsh Farm acting in effect as an extension of McGann's 'romantic ideology' in its suggestion of a Golden Age in which humankind lived in closer consort with nature.[3] Similarly, for Bate, the depiction of Cumbrian shepherds in book eight of *The Prelude* 'is not so much to show shepherds as they are but rather to bring forward an image of human greatness, to express faith in the perfectibility of mankind once institutions and hierarchies are removed and we are free, enfranchised, and in an unmediated, unalienated relationship with nature'.[4] Bate's ecocriticism is unafraid to assert that to be truly radical one must know how and what to conserve, and this particular recuperation of Wordsworth the nature poet risks association with a distinctly counter-revolutionary sense of literature as the search for timeless truths about the human condition. Again, in 'Poetry and Biodiversity', the search for a new politics, modelled on 'the order of nature', which would not at the same time be fascist, leads Bate, via 'the centrality to Green thinking of the notion of *conservation*', to propose Burkean conservatism.[5] The problem here, of which Bate is at some level distinctly aware, is what we mean when we speak of 'nature' and its 'order'.

My sketch of key ecocritical works thus far highlights the importance of romanticism as a reference point, organized around figures like Wordsworth and Thoreau, and this in turn focuses a number of key issues in the development of literary ecocriticism. First, at a general level, is it or should it be the function of ecocriticism to seek out ecologically sensitive works of literature and thus to constitute its own canon? And if so, would this consist exclusively of a 'nature writing' informed largely by the romantic tradition? These are questions to which I will return. More specific is the *kind* of ecological consciousness that a 'return' to romanticism might promote. In *Writing the Environment*, Gretchen Legler notes that 'the production and consumption of writing about nature' tends to depend upon 'nostalgia for a better-than-present world, a looking backward to a place and time not spoiled or polluted or industrialised'; N. H. Reeve, concluding an essay on the popular turn-of-the century nature writer W. H. Hudson, suggests that 'the romantic vision of nature will always include a sense that losing something may be preferable to finding it, since in loss it remains unexplorable and thus truly "protected".'[6]

Bate himself, aware of the associations of romanticism with escapism, invokes Freud's analogy of mental phantasy with the creation of nature-parks or 'reservations', each 'reclaimed from the encroaches of the reality-principle';[7] his attempt to offset this view with Hazlitt's theory that nature always implies 'recollection', and can therefore function as a universal and democratically structured 'home', does not necessarily disavow the Freudian suggestion.

If forgetting is integral to the psychic economy of romanticism, and if romanticism can exert such a powerful influence both on nature writing and on the formation of ecocriticism, there is growing evidence that contemporary ecocritics are troubled by the situation. In reviewing the recent classic of American environmentalism, Edward Abbey's *Desert Solitaire: A Season in the Wilderness* (1968), SueEllen Campbell arrives at an arresting conclusion: 'It is increasingly clear to me that environmental literature in general, and Abbey's book as an example, works partly by shutting out social and cultural complexities – an omission that's probably one source of the desire they embody and evoke.'[8] Campbell even borrows Michael Rogin's phrase 'political amnesia' to describe this process. Borne out of the immediate political complexities of the environmental crisis, ecocriticism presumably cannot afford to identify itself with the potentially complacent or reductive scenarios of its prehistory. It thus finds itself in a curiously ambivalent position with regard to the development of critical and cultural theory within the discipline of literary studies. One critic in particular, Dominic Head, has worried productively and engagingly over this relationship in two closely aligned essays which are important introductions to the contemporary debate.[9] On the one hand, as Head argues, 'the Green movement in general is predicated on a typically postmodernist deprivileging of the human subject'; alongside post-structuralism's philosophical critique of anthropocentrism, postcolonialist criticism is involved, in Head's phrase, in a process of 'decentring' and 'qualified recentring', unpicking existent hierarchies of value and venturing alternative priorities which, in the case of ecocriticism, would inevitably revolve around the concept of the planet-as-limit. There would thus appear to be many potentially fertile connections between literary ecocriticism and contemporary literary theory.

On the other hand, the rise of theory has arguably taken the academic study of literature to new levels of specialism. Head notes that 'the process of developed specialised thinking about language and literature may be self-serving, channelled in the direction of a contained professionalism'.[10] Perhaps more than ever, literary-critical discourses are the possession of knowledge-workers, who might once have been called elites, raising the spectre of precisely the kind of instrumental rationality which it is the *raison d'être* of Green movements to contest. In *Romantic Ecology*, Bate targets the critical position that '"there is no nature except as it is constituted by acts of political definition made possible by particular forms of government"',[11] reminding us of how far 'nature' became a scare term in the theoretical contexts of the 1980s, to be invoked only with quotation marks firmly clamped to its ears, and in evidence of the way in which ideology functions through an inversion of the cultural. In his analysis of how the novel might be read by ecocritics, Dominic Head affirms the complex '*inscription*' of social history and sexual politics in the fictional landscapes of Hardy and Lawrence; but he also warns of how an overriding emphasis on 'textuality' in literary studies can be interpreted as a sign of alienation or 'crisis of disconnection from the non-human Other', offering anti-theorists the opportunity to suggest connections between 'systems of eco-damage' and the generally assumed gap between world and text.[12]

The best recent ecocriticism thus emerges as a mode of highly self-conscious critical reflection rather than as canon-formation or the privileging of 'nature writing' as such. Theoretical subtlety and mobility delicately counterbalance a distrust of professionalism and the kind of clear subject–object distinction across which the exploitative critical gaze can operate: SueEllen Campbell tries to keep her 'peripheral vision sharp' and treasures her status as 'amateur and sampler'.[13] Equally crucial is the relation between a critical vigilance concerning 'nature' as a perpetually contested historical concept, and the need to instil a consciousness of the material world and its limits. Ecological movements cannot hope to keep pace with debates around, for example, gene technology without the former; the challenge is to ensure that a critique of essentialism does not collapse into a conviction that physical limits cannot be acknowledged or audited. In literary criticism, some such striving for balance is evident in Anne Fernihough's recent, fascinating rereading of D. H. Lawrence. While acknowledging that Lawrence's aesthetics take him perilously close to a Spenglerian totalitarianism, Fernihough's concern to dissociate Lawrence from fascism hinges on a close critical attention to variations in the concept of organicism. Such attention, she argues, is often lacking when deconstructive criticism turns its attention to the 'organic'; the 'constructedness' or metaphoricity of the category is exposed, but 'only to replace one logocentrism with another, namely the belief that the organic *always* and *necessarily* claims an essential link between language and world', thus constituting 'a stable point of reference against which the deconstructive enterprise can define itself'.[14] In associating Lawrence with a more radical history of the organic, Fernihough is able to posit a 'fractured organic' and an 'anti-imperialist aesthetics', Lawrence's work 'splintering the organic into an "endless confusion of differences"' which simultaneously also ties in with an affirmation of the material world.

Fernihough's work shows ecocriticism operating as a fundamental, informing perspective. The same sophistication surrounding 'nature' and the 'organic' promises to take ecocriticism forward, in a more consciously programmatic sense, in the writing of American critics such as Gretchen Legler and Patrick D. Murphy. Drawing on Donna Haraway's 'Cyborg Manifesto', Legler embraces the work of writers Mary Oliver, Joy Harjo and Lucille Clifton, who explore a conception of the politicized (raced, classed and sexed) cyborg body poised between nature and culture, belonging to neither and disrupting the binary itself. Legler is unequivocal in her contention that only through such 'body politics' can we move beyond an understanding of nature and nature writing rooted in the nostalgic and romantic, 'those two nearly useless positions which serve largely to freeze aesthetic and intellectual progress'.[15] Murphy appropriates Bakhtinian dialogics as the basis of an ecofeminist critical paradigm which is, at were, called into existence by new modes of literary expression: 'Existing paradigms do not seem to encompass the range of environmental literature that has been written and is being written today, nor for critiquing the diversity of expressions of human–non-human relationships, of the generation of geopsyche, or of the ecosystemic situatedness to be found in contemporary literature.'[16]

What we find in both critics is, of course, a sense that ecocriticism as a way of reading texts, and as the tracing of a new literary tradition, are not mutually exclusive developments. The urgencies of the environmental crisis have indeed brought forth a whole, primarily literary response even as they have brought forth literary ecocriticism. For Richard Kerridge, the crisis is inherently cultural, because it is a crisis of representation.[17] Like the radioactive particles of Chernobyl, environmental issues are either unthinkably vast or easily peripheralized; either way, they become invisible, unrepresentable. Kerridge invokes Slavoj Zizek's Lacanian account of the three modes by which the crisis is repressed: the refusal to take it seriously, because of the threat it poses to our conception of a well-regulated 'nature'; the obverse and obsessive response of the panic-driven activist; and the moral and teleological response by which impending global disaster is understood as a punishment for human wrong-doing. All hinder an appreciation of the 'reality', but also because 'the real' in this case simply cannot be contained by the symbolic order. In the light of Kerridge's approach, the work of Don DeLillo – in no immediate sense a pastoral writer – can appear central to the contemporary ecological movement in its restless, ongoing concern with the things we produce but cannot dispel. Nick Shay, the nuclear waste executive in *Underworld*, speaks of 'the kind of desperate crisis, the intractability of waste, that doesn't really seem to be taking place except in the conference reports and the newspapers. It is not otherwise touchable somehow, for all the menacing heft and breadth of the material, the actual pulsing thing.'[18] Ecocriticism seems impelled by the belief that what imaginative literature can offer is a mode of relatedness to, or a way of getting in touch with, a world we have developed complex ways of forgetting.

Notes

1 R. Kerridge, 'Introduction', in R. Kerridge and N. Sammels (eds), *Writing the Environment: Ecocriticism and Literature* (London, 1998), 5.
2 J. Bate, *Romantic Ecology: Wordsworth and the Environmental Tradition* (London, 1991), 8–9.
3 G. Holderness, *D. H. Lawrence: History, Ideology and Fiction* (Dublin, 1982), 182.
4 Bate, *Romantic Ideology*, 29.

5 J. Bate, 'Poetry and Biodiversity', in Kerridge and Sammels (eds), *Writing the Environment*, 53–70.
6 G. Legler, 'Body Politics in American Nature Writing. "Who may contest for what the body of nature will be?"', in Kerridge and Sammels (eds), *Writing the Environment*, 72; N. H. Reeve, 'Feathered Women: W. H. Hudson's *Green Mansions*', in Kerridge and Sammels (eds), *Writing the Environment*, 144.
7 S. Freud, 23rd 'Introductory Lecture on Psycho-Analysis'; quoted in Bate, *Romantic Ecology*, 52.
8 S. Campbell, 'Magpie', in Kerridge and Sammels (eds), *Writing the Environment*, 24.
9 D. Head, 'Problems in Ecocriticism and the Novel', in *Key Words: A Journal of Cultural Materialism*, 1 (1998), 60–73; 'The (Im)possibility of Ecocriticism', in Kerridge and Sammels (eds), *Writing the Environment*, 27–39.
10 Head, 'The (Im)possibility of ecocriticism', 29.
11 A. Liu, *Wordsworth: The Sense of History* (Stanford, 1989), quoted in Bate, *Romantic Ecology*, 15.
12 Head, 'Problems in Ecocriticism and the Novel', 65–6.
13 Campbell, 'Magpie', 14.
14 A. Fernihough, *D. H. Lawrence: Aesthetics and Ideology* (Oxford, 1993), 17.
15 Legler, 'Body Politics in American Nature Writing', 71.
16 P. D. Murphy, 'Anotherness and Inhabitation in Recent Multicultural American Literature', in Kerridge and Sammels (eds), *Writing the Environment*, 43. See also *Literature, Nature, and Other: Ecofeminist Critiques* (New York, 1995).
17 Kerridge, 'Introduction', 1–9.
18 D. DeLillo, *Underworld* (New York, 1997; London, 1998), 805.

Archetypal history: Simon Schama's *Landscape and Memory*

Bill Schwarz

There are few historians in Britain who could regard themselves as celebrities. Simon Schama, resident in North America, comes close. His books arrive with all the hype one has come to expect from any other potential bestseller. If not subject to the blandishments of *Hello!* magazine, he is at least up to receiving transatlantic envoys from *Modern Painters*, whose cerebral prose can't quite avoid the breathlessness required when introducing the celebrity and communicating the aura of his domestic world. ('The Schama residence was built in 1970 by a successful producer of television commercials who liked wide open spaces inside as well as out. Perched on a small ridge, it has a marvellous view over the lower Hudson valley – a sweeping vista of deciduous woodland stretching away as far as the eye can see.')[1] Publication of each volume, necessarily in such a situation, becomes something of an event, and expectations intensify: a little more iconoclastic, a little more literary, a little more off-beat. Barbara Cartland can write the same book a hundred times over and know that her readership will still be there. Those in more volatile sectors of the publishing market don't have that privilege. The pressures are real; it's daft to assume that those talented historians who have become hot commodities remain untouched.

Landscape and Memory was published in 1995, with all the attendant hoopla. It is a serious book, at times quite wonderful, which repays rereading. But it has its curiosities. Its three substantive sections are entitled 'Wood', 'Water' and 'Rock', carrying a faint echo of the kids' playground game of stone–paper–scissors. For the organizing concepts of a historical inquiry, these are provocatively abstract. Even so, in keeping with the spirit of such an exploration, I want briefly to respond to Schama's provocation by raising four issues: 'Roots', 'Magic', 'Narrative' and 'Plumb'.

Roots

The Jewish diaspora frames Schama's account, both in the narrative and analytically. Schama himself is a child of the diaspora, starting out life on the northern bank of the Thames estuary, in Leigh-on-Sea and then, when his family fortunes entered a periodic cyclical downturn, relocating to Golders Green. The prologue and opening chapter carry us back a couple of generations to (what has now become) the Polish-Lithuanian border and to (what once was) the mighty forest of Bialowieza. Their fine prose transports Schama and us to the dark shadowlands of the heart of Central Europe. 'The fields and forests and rivers had seen war and terror, elation and desperation; death and resurrection; Lithuanian kings and Teutonic knights, partisans and Jews; Nazi Gestapo and Stalinist NKVD. It is haunted land where greatcoat buttons from six generations of fallen soldiers can be discovered lying amidst the woodland ferns.'[2] From the oblique perspective of the more settled contours of the English landscape, in Poland there appears to be a tragic convergence of history and geography, in which every feature of the landscape is heavy with meaning. The destruction of the Polish state in the nineteenth century meant that the Polish nation was lived peculiarly intensely in the imagination, patriotic longing *located* in what was most immediate, sensuous and present. When the nation state reappeared in the twentieth century its citizens awoke, on more than one occasion, to find their nation had been moved in the night – a little to the east, a little to the west, but at every moment subject to the imperatives of external forces more powerful than itself. Landscapes and territories were not only steeped in the past; they were peculiarly politicized.

Yet Schama's story is concerned not only with the dislocations of the Poles but with the Jews who inhabited this larger national culture, simultaneously both inside and out. For all their cultivation, Polish Jewry was doubly displaced, creating imagined pasts – and their requisite landscapes – which truly were overdetermined. He quotes the famous aphorism of the distinguished Polish Jew Isaac Deutscher: 'Trees have roots, Jews have legs.'[3] As Schama himself discovered, confronted by the murderous histories of the past burnt into the landscapes of the present, those who have legs have the capacity to turn and walk away. In this sense, he implies, we can become our own agents of history: not forgetful, but neither consumed by the imperatives (in Melanie Klein's words) of 'manic reparation'.[4]

The Bialowieza forest, in this account, functions as the *mise-en-scène* of the great dramas of twentieth-century history in Central Europe – not merely backdrop, but active *historical* presence. There are some startling passages which re-create with imaginative power the contending political forces which have met, through the ages, on this terrain. The arrival of Hermann Goering is told in all its baroque horror. So too Schama recounts the extraordinary events surrounding the Nazi endeavour to seize the original manuscript by Tacitus, establishing – once and for all, in the eyes of Nazi ideologues – the transcendent origin of the German people. In this lies the manic compulsion to think with the blood – to adhere to an absolutist conception of ethnicity which is implacable in segregating those who belong – and those who don't. This is, as we know, a kind of history which is driven by a faith in genetics, or by monologic conceptions of cultural purity, in which the past determines unilaterally. The act of walking – choosing to turn away; to cross frontiers; to walk towards and embrace others is, in these imaginative scenarios, simply not possible. Merely to take the first step is to cross the threshold of treason to the *volk*.

To scrape away the layers of the present so that buried histories come alive is, in a sense, what a vital historiography is about, allowing us – precisely – not to be forgetful. The six generations of fallen soldiers become, thanks to Schama's careful archaeology, resurrected. This is Schama's modest commemoration of the anonymous footsoldiers of the past. Commemoration, he suggests in *Citizens*, his history of the French Revolution, is most appropriate 'when least monumental'.[5]

Yet this is also where, in *Landscape and Memory*, the abstraction and the archetypes intervene. It is not only six generations that Schama wishes to resurrect, but some primal memory – signified by the forest, by 'Wood' itself – which lies present in our civilization of today, but buried and unseen. The tone of his text suggests that the Bialowieza forest stands in unspecified proximity to the origins of Europe. It signifies 'the primary bedrock' of an essential element of European civilization which, as a result of the labours of those historians with the eyes to see, can be recovered again, carried 'toward the light of contemporary recognition'.[6] Whether the series of case-studies in the book persuades the reader is, evidently, the key issue. I remain sceptical. But my scepticism derives, in part, from doubts about the endeavour itself.

Schama says precious little about the nature of the primal forms he identifies – except to distance himself, quietly, from Jung, whom he regards as radically ahistorical. We know nothing analytically about these archetypes. Nor, in the book's most serious collapse, does he ever explain what he means either by landscape or by memory – save his recourse to the catch-all (and pretty unhelpful) abstraction of 'social memory'. To say this might seem to invite Schama to undertake a form of conceptual exegesis which he gives every indication of deploring, believing it results in nothing more than a tedious, vacuous academicism. One can see his point. But the absence of such clarification allows an unintended narrative to emerge, which is at odds with what I understand to be the formal argument of the book. Memory, here, moves for the most part along no visible channels: it appears to have no localized histories. It is atavistic, jumping out of the primal caves of darkest, ancient Europe, carried forward through time we know not how, and by some means comes to be present in the unknown recesses of the cultivated mind of contemporary civilization. Such a conception hovers close to a whole spectrum of older, and frankly dangerous, ideas about the continuities of racial memory. There is something too unilateral about such a view. It summons, I think, less history than an unrequited longing – a longing, after all, for roots.

Magic

Perhaps such longing, inflected in this way, is testimony to the peculiarities of the English diaspora. When his father reflected on mortality, we are told, he drew upon an idiom drawn from the language of cricket. And there is too a knowing confirmation of national affiliation in Schama opening the volume by declaring his affection for Kipling. 'For a small boy with his head in the past, Kipling's fantasy was potent magic.'[7] The primal forms or archetypes identified by Schama retain – in the modern epoch – their mythic or magical capacities. *Landscape and Memory*, in so far as it proselytizes at all, is a book which tries to allow us to see in new ways; or rather, in more existential mode, this is what the book tries itself to *be* – a new way of seeing. The argument which follows is one which is pitted against (one reading of) the classical theorists of modernity. The iron cage of modernity, and its attendant disenchantments, do not compose for Schama the full story. 'For if the entire history of landscape in the West is indeed just a mindless race toward a machine-driven universe, uncomplicated by myth, metaphor, and allegory, where measurement, not memory, is the absolute arbiter of value, where our ingenuity is our tragedy, then we are indeed trapped in the engine of our self-destruction.'[8] He is at his most passionate when he declares his love for the enchantments of the modern world. Modernity, he argues, has done nothing to forestall the vitality of the mythic and the magical.

This is a welcome intervention – especially at a time when there exists an entire historiography influenced by a kind of pop Foucauldianism in which, precisely, ingenuity transmogrifies implacably into the will for power. Underwriting Schama's arguments is the unstated belief that it is through the aesthetic imagination – through the work of painters, poets and visionary historians – that this mythic world can be revealed. The evidence for the bulk of the book is drawn from such sources – testament, perhaps, to a Romantic or strangely skewed idea of the humanities. Given Schama's declared populist sympathies (evident, once more, in his reminiscences about the social organization of English cricket) it is odd that he has drawn so selectively from the dominating forms of contemporary mass culture.[9] Most puzzling of all is the absence of the cinema. Where else, in the twentieth century, have citizens imbibed so deeply the invention of landscape *but* through the magic of cinema? This lack of curiosity about the modern vernacular is one difficulty. Another, more pressing to my mind, is the belief that it is the job (or a job) of the artist to recover the hidden connections to a past which is essentially primitive. One can see why the disclaimer about Jung was necessary. Schama is a million miles away from Freud at this point, as he would be the first to admit, where the idea of the primitive can signify a complex, dynamic and contemporary psychic process. The disclaimer notwithstanding, he is as close as could be to Jung and his archetypes. The task of bridging such a starting point to historical inquiry is indeed awkward. How are the connections to be plotted?

Narrative

Despite the multiplicity of contending explanations of what is happening in historiography today, one thing is clear: an inherited academic positivism is on the wane. In its place, commitment to a poetics of historical interpretation is becoming commonplace. Such a move places great emphasis on the semiotic as a method, and greater interest in those subjects which were eclipsed by the monologic eye of an overly rationalistic positivism. Yet it isn't at all clear what are the appropriate narrative forms which follow.

Schama is interesting in this respect. He has a reputation for being a leading exponent of the new cultural history, inventing narrative forms which unsettle the norms of academic expectation. Arguably, the anthropological organization of his study of sixteenth-century Dutch culture, *The Embarrassment of Riches*, owes more to the structuralist analysis of mentalities (or at least to Durkheim) than it does to conventional anglophone historiography.[10] *Dead Certainties*, experimental and consciously challenging academic historians' deepest conventions, coquettes with the idea of an open text, unhindered by the search for causal connections or by an empirically verifiable concluding explanation.[11] *Citizens* calls itself a 'chronicle', and adopts a deliberately old-fashioned chronological narrative – a challenge (in this instance) to his own perceptions of a desiccated social history and, indeed, to the sort of anthropological history which once he had made his own.

This willingness to experiment with narrative, and to think consciously of different narrative forms for different objects of historical inquiry, is a winning aspect of Schama's work. One never knows what will come next. Yet with more irony than self-deprecation he has suggested that it has been his incompetence with social history which has been determining, disclaiming all allegiance to the new: 'I was the tail end of the old narrative history not the cutting edge of the new.'[12]

At best, there is an imagistic method in Schama. This is particularly evident in his manner of opening his narratives. In *Embarrassment of Riches* he hangs an entire reading of early modern civilization in the Netherlands on his account of the drowning cell; in *Citizens* his account of the giant plaster elephant which once existed on the site of the Bastille serves to hold in the reader's mind the pressing question of the ways in which the French Revolution has been remembered; and in *Landscape and Memory* his reconstruction of the history of the Bialowieza forest dominates what is to come. But in *Landscape and Memory*, in particular, everything thereafter unravels. Schama calls this a narrative which 'meanders'. To readers less indulgent than he, it looks entirely random, narrative structure giving way to descriptive excess.

One can see the problem. Above all, Schama positions himself as the enemy of dispassionate history. With the ghost of Richard Cobb haunting him, he is always keen to walk, to see, to touch: there is something of the *flaneur* about them both. There is nothing he likes better than to tell a good story. His narrative models, in these terms, cannot be found in the contemporary academy, but rather in the writing of historians of public repute who were happily free of the usual constraints of academic history. Macaulay and Trevelyan, as supremely English and magical in their way as Kipling, are present for Schama in his imaginative act of revitalizing history. At the same time however, as his concern with archetypes suggests, he is also interested intellectually in how the human mind works: with the emotional, psychic and aesthetic properties of memory and (in *Landscape and Memory*) with the connections holding together the modern and the primitive. But without some guiding constraints of form narrative truly does become random, moving from one well-crafted evocation to the next. This isn't to suggest that Schama immerses himself in the formalism of Saussure, Freud or Lévi-Strauss – although in ambition, if we take his words seriously, he is not far removed from them (or, in another mode, from the Carlo Ginzburg of *Ecstasies*).[13] But deep in his work there rests a false dichotomy, in which abstraction is the nerveless posturing of academics who have abdicated themselves from the concerns of real life, while the true heroes of intellectual life are the impassioned storytellers who remain close to the people. This is a tired old myth. *Passion* is not the issue.

Plumb

Schama comes from a discernible intellectual heritage of historians. There is a connection, through the Cambridge historian J. H. Plumb, back to Trevelyan and Macaulay. Schama himself is one of a constellation of talented historians who came under Plumb's influence and who now loom large in the public eye: one could mention, among others, David Cannadine, Linda Colley, Peter Hennessey. Plumb's credo has turned on a patrician iconoclasm, exuding an ironic scepticism which derives from one who has been near the centre of things for much of his life. There is here an irreverence for the way things are, which can masquerade as a kind of radicalism, while at the same time carrying a disdain for the ordinary and for the run-of-the-mill. Plumb's influence on generating a prominent group of historians in England, very conscious of its intellectual antecedents in the national culture, has been impressive.[14]

Schama is palpably his own man. In reasserting the mythic and primitive as constituents of the contemporary, and in welcoming the capacity of these prerational forms still to enchant daily life, he distanced himself from Plumb's renowned, and still engaging, polemic announcing 'the death of the past'.[15] But the legacy – acknowledged generously by Schama – is clear enough.

It is, though, complex. There is a residual populism present, in which the people are championed against the bureaucratic and the technocratic. There is a truly ambivalent regard for the ordinary, which (in professional terms) is merely humdrum, but (in historiographic terms) is that which is positioned against an overly cognitive, or messianic, politics, in which 'ordinary people' are recruited to the cause of History. And there is an immovable disdain for intellectual systems, which can only deaden thought and replace life by abstraction, and which gridlocks the chaos and contingencies of history. In place of systematization – in predictably English mode – comes a particular recourse to a poetics of history: literary, descriptive and imaginative. But for these to be plucked, as if from thin air, as innocent terms – without any prior conceptual baggage – is a sign of a surprisingly careless attitude to ideas. The debt to a Romantic sensibility is significant. It is not, in my view, that there is anything worrying about historical explanation which is literary, descriptive or 'imaginative'. It is simply that if these terms are to serve as an escape-hatch from the rigor mortis of academic history, their consequences need to be thought through. This spontaneous recourse to inherited English traditions is not without its complications. After all, part of the intellectual consequences of modernism – in aesthetics, but also in what now passes for 'theory' (psychoanalysis, structuralism, literary formalism and so on) – worked to elaborate some of the central issues to do with the imagination, representation and narrative. To go back, by default as it were, to prior intellectual formations, as if nothing much had happened, is to miss too much.

For all its championing of Anselm Kiefer, *Landscape and Memory* remains peculiarly antiquarian, the dominating rhythms of twentieth-century culture passing it by. One need only contrast Schama's recuperation of the mythic to the renowned intervention by Barthes, some four decades ago, to understand mass culture as myth, in *Mythologies*.[16] There are still many fruitful avenues to be followed arising from this, on which Schama is silent. But the connections are there. Barthes' *Mythologies* followed directly from his study of Michelet (a defining presence in Schama's *Citizens*), the one an elaboration of the other. *Mythologies* remains an important exploration of the historical imagination. At the end of his life Michelet claimed that 'I was born of the people, I have the people in my heart', but went on to lament: 'But the people's language, its language was inaccessible to me. I have not been able to make the people speak.'[17] It is such sentiment which made Barthes reflect on the role of historian as 'magus', seeing

> the historian [as] the man who has reversed Time, who turns back to the place of the dead and recommences their life in a clear and useful direc-

> tion; he is the demiurge who links what was scattered, discontinuous, incomprehensible: he weaves together the threads of all lives, he knits up the great fraternity of the dead, whose formidable displacement, through Time, forms that extension of History which the historian leads while walking backwards, gathered within his gaze which decides and discloses.[18]

How could Schama resist?

Notes

1 P. Wright, 'Schama's Country. An Interview with Simon Schama', *Modern Painters* 8(1) (1995), 10.
2 S. Schama, *Landscape and Memory* (London, 1995), 24.
3 Schama, *Landscape*, 29.
4 M. Klein, *Love, Guilt and Reparation* (London, 1975), 278.
5 S. Schama, *Citizens: A Chronicle of the French Revolution* (Harmondsworth, 1989), 5.
6 Schama, *Landscape*, 17.
7 Schama, *Landscape*, 3.
8 Schama, *Landscape*, 14.
9 Wright, 'Schama Country', 13.
10 S. Schama, *The Embarrassment of Riches: An Interpretation of Dutch Culture in the Golden Age* (London, 1991).
11 S. Schama, *Dead Certainties (Unwarranted Speculations)* (London, 1991).
12 Wright, 'Schama Country', 11.
13 C. Ginzburg, *Ecstasies: Deciphering the Witches' Sabbath* (London, 1990).
14 D. Cannadine, *G. M. Trevelyan* (London, 1992).
15 J. H. Plumb, *The Death of the Past* (London, 1969).
16 R. Barthes, *Mythologies* (London, 1986).
17 R. Barthes, *Michelet* (New York, 1987), 199.
18 Barthes, *Michelet*, 82–3.

NAME INDEX

SUBJECT INDEX